Catalan Numbers

Catalan numbers are probably the most ubiquitous sequence of numbers in mathematics. This book provides, for the first time, a comprehensive collection of their properties and applications in combinatorics, algebra, analysis, number theory, probability theory, geometry, topology, and other areas.

After an introduction to the basic properties of Catalan numbers, the book presents 214 different kinds of objects that are counted using Catalan numbers, in the form of exercises with solutions. The reader can try solving the exercises or simply browse through them. An additional sixty-eight exercises with prescribed difficulty levels present various properties of Catalan numbers and related numbers, such as Fuss-Catalan numbers, Motzkin numbers, Schröder numbers, Narayana numbers, super Catalan numbers, q-Catalan numbers, and (q, t)-Catalan numbers. The book concludes with a history of Catalan numbers by Igor Pak and a glossary of key terms.

Whether your interest in mathematics is recreation or research, you will find plenty of fascinating and stimulating facts here.

RICHARD P. STANLEY is Professor of Applied Mathematics at the Massachusetts Institute of Technology. He is universally recognized as a leading expert in the field of combinatorics and its applications in a variety of other mathematical disciplines. He won the 2001 AMS Leroy P. Steele Prize for Mathematical Exposition for his books *Enumerative Combinatorics: Volume 1* and *Volume 2*, which contain material that form the basis for much of the present book.

T0331582

Catalan Numbers

RICHARD P. STANLEY

Massachusetts Institute of Technology

CAMBRIDGE
UNIVERSITY PRESS

CAMBRIDGE
UNIVERSITY PRESS

32 Avenue of the Americas, New York NY 10013-2473, USA

Cambridge University Press is part of the University of Cambridge.

It furthers the University's mission by disseminating knowledge in the pursuit of education, learning and research at the highest international levels of excellence.

www.cambridge.org
Information on this title: www.cambridge.org/9781107075092

© Cambridge University Press 2015

First published 2015

A catalogue record for this publication is available from the British Library

Library of Congress Cataloguing in Publication data
Stanley, Richard P., 1944–
Catalan numbers / Richard P. Stanley, Massachusetts Institute of Technology.
pages   cm
Includes bibliographical references and index.
ISBN 978-1-107-07509-2 (hardback) – ISBN 978-1-107-42774-7 (pbk.)
1. Catalan numbers (Mathematics)   2. Combinatorial analysis.
3. Matrices.   I. Title.
QA164.S74   2015
511'.6–dc23   2014043732

ISBN 978-1-107-07509-2 Hardback
ISBN 978-1-107-42774-7 Paperback

Contents

Preface

This text had its origins in the 1970s, when I first started teaching enumerative combinatorics and became aware of the ubiquity of Catalan numbers. Originally I just made a handwritten list for my own benefit. One of the earliest such lists has survived the ravages of time and appears in Appendix A. Over the years, the list became larger and more sophisticated. When I wrote the second volume of *Enumerative Combinatorics* (published in 1999), I included sixty-six combinatorial interpretations of Catalan numbers (Exercise 6.19) as well as numerous other exercises related to Catalan numbers. Since then I have continued to collect information on Catalan numbers, posting most of it on my "Catalan addendum" web page. Now the time has come to wrap up this 40+ years of compiling Catalan material, hence the present monograph. Much of it should be accessible to mathematically talented undergraduates or even high school students, while some parts will be of interest primarily to research mathematicians.

This monograph centers on 214 combinatorial interpretations of Catalan numbers (Chapters 2 and 3). Naturally some subjectivity is involved in deciding what should count as a new interpretation. It would be easy to expand the list by several hundred more entries by a little tweaking of the current items or by "transferring bijections." For instance, there is a simple bijection φ between plane trees and ballot sequences. Thus, whenever we have a description of a Catalan object in terms of plane trees, we can apply φ and obtain a description in terms of ballot sequences. I have used my own personal tastes in deciding which such descriptions are worthwhile to include. If the reader feels that 214 is too low a number, then he or she can take solace in the solution to item 65, which discusses *infinitely* many combinatorial interpretations.

Also central to this monograph are the sixty-eight additional problems related to Catalan numbers in Chapters 4 and 5. Some of these problems deal with generalizations, refinements, and variants of Catalan numbers, namely, q-Catalan numbers, (q, t)-Catalan numbers, (a, b)-Catalan numbers, Fuss-Catalan numbers, super Catalan numbers, Narayana numbers, Motzkin numbers, and Schröder numbers. Here we have made no attempt to be as comprehensive as in Chapters 2 and 3, but we hope we have included enough information to convey the flavor of these objects.

In order to make this monograph more self-contained, we have included a chapter on basic properties of Catalan numbers (Chapter 1) and a Glossary. The Glossary defines many terms that in the text are simply given as citations to the two volumes of *Enumerative Combinatorics*. (See references [64] and [65] in the bibliography.)

The history of Catalan numbers and their ilk is quite interesting, and I am grateful to Igor Pak for contributing Appendix B on this subject. There the reader can find much fascinating information on a subject that has not hitherto received adequate attention.

Innumerable people have contributed to this monograph by sending me information on Catalan numbers. They are mentioned in the relevant places in the text. Special thanks go to David Callan, Emeric Deutsch, Igor Pak, and Lou Shapiro for their many contributions. The author was partially supported by the National Science Foundation.

1

Basic Properties

1.1. The Definition of Catalan Numbers

There are many equivalent ways to define Catalan numbers. In fact, the main focus of this monograph is the myriad combinatorial interpretations of Catalan numbers. We also discuss some algebraic interpretations and additional aspects of Catalan numbers. We choose as our basic definition their historically first combinatorial interpretation, whose history is discussed in Appendix B along with further interesting historical information on Catalan numbers. Let $\mathcal{P}_{n+2}$ denote a convex polygon in the plane with $n+2$ vertices (or *convex $(n+2)$-gon* for short). A *triangulation* of $\mathcal{P}_{n+2}$ is a set of $d-1$ diagonals of $\mathcal{P}_{n+2}$ which do not cross in their interiors. It follows easily that these diagonals partition the interior of $\mathcal{P}_{n+2}$ into n triangles. Define the nth *Catalan number* C_n to be the number of triangulations of $\mathcal{P}_{n+2}$. Set $C_0 = 1$. Figure 1.1 shows that $C_1 = 1$, $C_2 = 2$, $C_3 = 5$, and $C_4 = 14$. Some further values are $C_5 = 42$, $C_6 = 132$, $C_7 = 429$, $C_8 = 1430$, $C_9 = 4862$, and $C_{10} = 16796$.

In this chapter we deal with the following basic properties of Catalan numbers: (1) the fundamental recurrence relation, (2) the generating function, (3) an explicit formula, and (4) the primary combinatorial interpretations of Catalan numbers. Throughout this monograph we use the following notation:

$\mathbb{C}$	complex numbers
$\mathbb{R}$	real numbers
$\mathbb{Q}$	rational numbers
$\mathbb{Z}$	integers
$\mathbb{N}$	nonnegative integers $\{0, 1, 2, \dots\}$
$\mathbb{P}$	positive integers $\{1, 2, 3, \dots\}$
$[n]$	the set $\{1, 2, \dots, n\}$, where $n \in \mathbb{N}$
$\#S$	number of elements of the (finite) set S

1

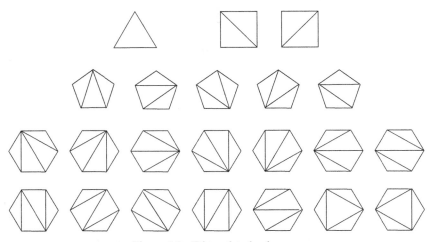

Figure 1.1. Triangulated polygons.

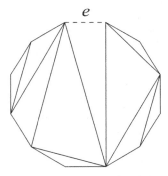

Figure 1.2. The recursive structure of a triangulated polygon.

1.2. The Fundamental Recurrence

To obtain a recurrence relation for Catalan numbers, let $\mathcal{P}_{n+3}$ be a convex $(n+3)$-gon. Fix an edge e of $\mathcal{P}_{n+3}$, and let T be a triangulation of $\mathcal{P}_{n+3}$. When we remove the edge e from T we obtain two triangulated polygons, say $\mathcal{Q}_1$ and $\mathcal{Q}_2$, in counterclockwise order from e, with one common vertex. If $\mathcal{Q}_i$ has $a_i + 2$ vertices, then $a_1 + a_2 = n$. See Figure 1.2, where $n = 9$, $a_1 = 5$, and $a_2 = 4$. It is possible that one of $\mathcal{Q}_1$ or $\mathcal{Q}_2$ is just a single edge, which occurs when the triangle of T containing e has an additional edge on $\mathcal{P}_{n+3}$, necessarily adjacent to e. In this case we consider the edge as a 2-gon, which has $C_0 = 1$ triangulations.

Conversely, given two triangulated polygons with $a_1 + 2$ and $a_2 + 2$ vertices, we can put them together to form a triangulated $(n + 3)$-gon by reversing the above procedure. Since there are C_{a_i} triangulations of $\mathcal{Q}_i$, we obtain the recurrence and initial condition

$$C_{n+1} = \sum_{k=0}^{n} C_k C_{n-k}, \quad C_0 = 1. \tag{1.1}$$

This is the most important and most transparent recurrence satisfied by C_n. It easily explains many of the combinatorial interpretations of Catalan numbers, where the objects being counted have a decomposition into two parts, analogous to what we have just done for triangulated polygons. For other "Catalan objects," however, it can be quite difficult, if not almost impossible, to see directly why the recurrence (1.1) holds.

1.3. A Generating Function

Given the recurrence (1.1), it is a routine matter for those familiar with generating functions to obtain the next result. For some background information on generating functions and the rigorous justification for our manipulations, see [64], especially Chapter 1. Let us just mention now one aspect of generating functions, namely, the binomial theorem for arbitrary exponents. When a is any complex number, or even an indeterminate, and $k \in \mathbb{N}$, then we define the *binomial coefficient*

$$\binom{a}{k} = \frac{a(a-1)\cdots(a-k+1)}{k!}.$$

The "generalized binomial theorem" due to Isaac Newton asserts that

$$(1+x)^a = \sum_{n \geq 0} \binom{a}{n} x^n. \tag{1.2}$$

This formula is just the formula for the Taylor series of $(1+x)^a$ at $x = 0$. For our purposes we consider generating function formulas such as Equation (1.2) to be "formal" identities. Questions of convergence are ignored.

1.3.1 Proposition. *Let*

$$C(x) = \sum_{n \geq 0} C_n x^n$$

$$= 1 + x + 2x^2 + 5x^3 + 14x^4 + 42x^5 + 132x^6 + 429x^7 + 1430x^8 + \cdots.$$

Then

$$C(x) = \frac{1 - \sqrt{1 - 4x}}{2x}. \tag{1.3}$$

Proof. Multiply the recurrence (1.1) by x^n and sum on $n \geq 0$. On the left-hand side we get

$$\sum_{n \geq 0} C_{n+1} x^n = \frac{C(x) - 1}{x}.$$

Since the coefficient of x^n in $C(x)^2$ is $\sum_{k=0}^{n} C_k C_{n-k}$, on the right-hand side we get $C(x)^2$. Thus

$$\frac{C(x) - 1}{x} = C(x)^2,$$

or

$$xC(x)^2 - C(x) + 1 = 0. \tag{1.4}$$

Solving this quadratic equation for $C(x)$ gives

$$C(x) = \frac{1 \pm \sqrt{1 - 4x}}{2x}. \tag{1.5}$$

We have to determine the correct sign. Now, by the binomial theorem for the exponent $1/2$ (or by other methods),

$$\sqrt{1 - 4x} = 1 - 2x + \cdots.$$

If we take the plus sign in Equation (1.5) we get

$$\frac{1 + (1 - 2x + \cdots)}{2x} = \frac{1}{x} - 1 + \cdots,$$

which is not correct. Hence we must take the minus sign. As a check,

$$\frac{1 - (1 - 2x + \cdots)}{2x} = 1 + \cdots,$$

as desired. $\square$

1.4. An Explicit Formula

From the generating function it is easy to obtain a formula for C_n.

1.4.1 Theorem. *We have*

$$C_n = \frac{1}{n+1} \binom{2n}{n} = \frac{(2n)!}{n! \, (n+1)!}. \tag{1.6}$$

Proof. We have

$$\sqrt{1-4x} = (1-4x)^{1/2} = \sum_{n \geq 0} \binom{1/2}{n} x^n.$$

Hence by Proposition 1.3.1,

$$C(x) = \frac{1}{2x} \left(1 - \sum_{n \geq 0} \binom{1/2}{n} (-4x)^n \right)$$

$$= -\frac{1}{2} \sum_{n \geq 0} \binom{1/2}{n+1} (-4)^{n+1} x^n.$$

Equating coefficients of x^n on both sides gives

$$C_n = -\frac{1}{2} \binom{1/2}{n+1} (-4)^{n+1}. \tag{1.7}$$

It is a routine matter to expand the right-hand side of Equation (1.7) and verify that it is equal to $\frac{1}{n+1} \binom{2n}{n}$. $\qquad\square$

The above proof of Theorem 1.4.1 is essentially the same as the proof of Bernoulli-Euler-Segner discussed in Appendix B. In Section 1.6 we will present a more elegant proof.

The expression $\frac{1}{n+1} \binom{2n}{n}$ is the standard way to write C_n explicitly. There is an equivalent expression that is sometimes more convenient:

$$C_n = \frac{1}{2n+1} \binom{2n+1}{n}. \tag{1.8}$$

Note also that

$$C_n = \frac{1}{n} \binom{2n}{n-1}.$$

1.5. Fundamental Combinatorial Interpretations

Among the myriad of combinatorial interpretations of Catalan numbers, a few stand out as being the most fundamental, namely, polygon triangulations (already considered), binary trees, plane trees, ballot sequences, parenthesizations, and Dyck paths. They will be the subject of the current section. For all of them the recurrence (1.1) is easy to see, or, what amounts to the same thing, there are simple bijections among them. We begin with the relevant definitions.

A *binary tree* is defined recursively as follows. The empty set $\emptyset$ is a binary tree. Otherwise a binary tree has a *root vertex* v, a *left subtree* T_1, and a *right*

Figure 1.3. The five binary trees with three vertices.

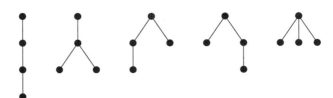

Figure 1.4. The five plane trees with four vertices.

subtree T_2, both of which are binary trees. We also call the root of T_1 (if T_1 is nonempty) the *left child* and the root of T_2 (if T_2 is nonempty) the *right child* of the vertex v. We draw a binary tree by putting the root vertex v at the top, the left subtree T_1 below and to the left of v, and the right subtree T_2 below and to the right of v, with an edge drawn from v to the root of each nonempty T_i. Figure 1.3 shows the five binary trees with three vertices.

A *plane tree* (also called an *ordered tree* or *Catalan tree*) P may be defined recursively as follows. One specially designated vertex v is called the *root* of P. Thus plane trees, unlike binary trees, cannot be empty. Then either P consists of the single vertex v, or else it has a sequence $(P_1, \ldots, P_m)$ of *subtrees* P_i, $1 \le i \le m$, each of which is a plane tree. Thus the subtrees of each vertex are linearly ordered. When drawing such trees, these subtrees are written in the order left-to-right. The root v is written on the top, with an edge drawn from v to the root of each of its subtrees. Figure 1.4 shows the five plane trees with four vertices.

A *ballot sequence* of length $2n$ is a sequence with n each of 1's and -1's such that every partial sum is nonnegative. The five ballot sequences of length six (abbreviating -1 by just $-$) are given by

$$111--- \qquad 11-1-- \qquad 11--1- \qquad 1-11-- \qquad 1-1-1-.$$

The term "ballot sequence" arises from the following scenario. Two candidates A and B are running in an election. There are $2n$ voters who vote sequentially for one of the two candidates. At the end each candidate receives n votes. What is the probability p_n that A never trails B in the voting and both candidates receive n votes? If we denote a vote for A by 1 and a vote for B by -1, then clearly the sequence of votes forms a ballot sequence if and only if A

never trails B. Moreover, the total number of ways in which the $2n$ voters can cast n votes for each of A and B is $\binom{2n}{n}$. Hence if f_n denotes the number of ballot sequences of length $2n$, then $p_n = f_n / \binom{2n}{n}$. We will see in Theorem 1.5.1(iv) that $f_n = C_n$, so $p_n = 1/(n+1)$. For the generalization where A receives m votes and B receives $n \le m$ votes, see Problem A2.[1] For the history behind this result, see Section B.7.

A *parenthesization* or *bracketing* of a string of $n+1$ x's consists of the insertion of n left parentheses and n right parentheses that define n binary operations on the string. An example for $n = 6$ is $(((xx)x)((xx)(xx)))$. In general we can omit the leftmost and rightmost parentheses without loss of information. Thus our example denotes the product of $(xx)x$ with $(xx)(xx)$, where $(xx)x$ denotes the product of xx and x, and $(xx)(xx)$ denotes the product of xx and xx. There are five ways to parenthesize a string of four x's, namely,

$$x(x(xx)) \quad x((xx)x) \quad (xx)(xx) \quad (x(xx))x \quad ((xx)x)x.$$

Let S be a subset of $\mathbb{Z}^d$. A *lattice path* in $\mathbb{Z}^d$ of length k with steps in S is a sequence $v_0, v_1, \ldots, v_k \in \mathbb{Z}^d$ such that each consecutive difference $v_i - v_{i-1}$ lies in S. We say that L *starts at* v_0 and *ends at* v_k, or more simply that L *goes from* v_0 *to* v_k. A *Dyck path* of length $2n$ is a lattice path in $\mathbb{Z}^2$ from $(0,0)$ to $(2n,0)$ with steps $(1,1)$ and $(1,-1)$, with the additional condition that the path never passes below the x-axis. Figure 1.5(a) shows the five Dyck paths of length six (so $n = 3$). A trivial but useful variant of Dyck paths (sometimes also called a Dyck path) is obtained by replacing the step $(1,1)$ with $(1,0)$ and $(1,-1)$ with $(0,1)$. In this case we obtain lattice paths from $(0,0)$ to (n,n) with steps $(1,0)$ and $(0,1)$, such that the path never rises above the line $y = x$. See Figure 1.5(b) for the case $n = 3$.

It will come as no surprise that the objects we have just defined are counted by Catalan numbers. We will give simple bijective proofs of this fact. By a *bijective proof* we mean a proof that two finite sets S and T have the same cardinality (number of elements) by exhibiting an explicit bijection $\varphi: S \to T$. To *prove* that φ is a bijection, we need to show that it is injective (one-to-one) and surjective (onto). This can either be shown directly or by defining an *inverse* $\psi: T \to S$ such that $\psi \varphi$ is the identity map on S, i.e., $\psi \varphi(x) = x$ for all $x \in S$, and $\varphi \psi$ is the identity map on T. (Note that we are composing functions *right-to-left*.) Often we will simply define φ without proving that it is a bijection when such a proof is straightforward. As we accumulate more and more objects counted by Catalan numbers, we have more and more choices

---

[1] A reference to a problem whose number is preceded by A refers to a problem in Chapter 4.

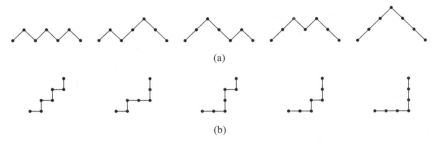

(a)

(b)

Figure 1.5. The five Dyck paths of length six.

for which of these objects we can biject to when trying to show that some new objects are also counted by Catalan numbers.

1.5.1 Theorem. *The Catalan number C_n counts the following:*

(i) *Triangulations T of a convex polygon with $n + 2$ vertices.*
(ii) *Binary trees B with n vertices.*
(iii) *Plane trees P with $n + 1$ vertices.*
(iv) *Ballot sequences of length 2n.*
(v) *Parenthesizations (or bracketings) of a string of $n + 1$ x's subject to a nonassociative binary operation.*
(vi) *Dyck paths of length 2n.*

Proof. (i)→(ii) (that is, the construction of a bijection from triangulations T of polygons to binary trees B). Fix an edge e of the polygon as in Figure 1.2. Put a vertex in the interior of each triangle of T. Let the root vertex v correspond to the triangle for which e is an edge. Draw an edge f' between any two vertices that are separated by a single edge f of T. As we move along edges from the root to reach some vertex v after crossing an edge f of T, we can traverse the edges of the triangle containing v in counterclockwise order beginning with the edge f. Denote by f_1 the first edge after f and by f_2 the second edge. Then we can define the (possible) edge f_1' crossing f_2 to be the *left edge* of the vertex v of B, and similarly f_2' is the *right edge*. Thus we obtain a binary tree B, and this correspondence is easily seen to be a bijection. Our rather long-winded description should become clear by considering the example of Figure 1.6(a), where the edges of the tree B are denoted by dashed lines. In Figure 1.6(b) we redraw B in "standard" form.

   (iii)→(ii) Given a plane tree P with $n + 1$ vertices, first remove the root vertex and all incident edges. Then remove every edge that is not the leftmost edge from a vertex. The remaining edges are the left edges in a binary tree B

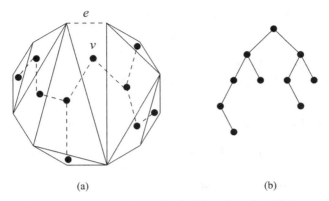

(a)   (b)

Figure 1.6. A binary tree associated with a triangulated polygon.

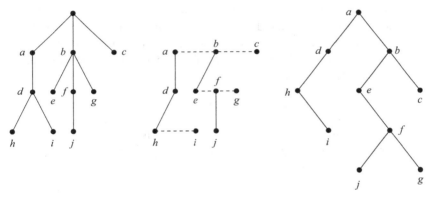

Figure 1.7. A binary tree constructed from a plane tree.

whose root is the leftmost child of the root of P. Now draw edges from each child w of a vertex v of P to the next child (the one immediately to the right of w) of v (if such a child exists). These horizontal edges are the right edges of B. The steps can be reversed to recover P from B, so the map $P \mapsto B$ gives the desired bijection. See Figure 1.7 for an example. On the left is the plane tree P. In the middle is the binary tree B with left edges shown by solid lines and right (horizontal) edges by dashed lines. On the right is B rotated 45° clockwise and "straightened out" so it appears in standard form.

This elegant bijection is due to de Bruijn and Morselt [11]. Knuth [33, §2.3.2] calls it the "natural correspondence." For some extensions, see Klarner [32].

(iii)→(iv) We wish to associate a ballot sequence $a_1 a_2 \cdots a_{2n}$ of length n with a plane tree P with $n + 1$ vertices. To do this, we first need to define a certain canonical linear ordering on the vertices of P, called *depth first order* or *preorder*, and denoted ord(P). It is defined recursively as follows.

Let $P_1, \ldots, P_m$ be the subtrees of the root v (listed in the order defining P as a plane tree). Set

$$\text{ord}(P) = v, \text{ord}(P_1), \ldots, \text{ord}(P_m) \text{ (concatenation of words)}.$$

The preorder on a plane tree has an alternative informal description as follows. Imagine that the edges of the tree are wooden sticks, and that a worm begins facing left just above the root and crawls along the outside of the sticks, until (s)he (or it) returns to the starting point. Then the order in which vertices are seen for the first time is preorder. Figure 1.8 shows the path of the worm on a plane tree P, with the vertices labeled 1 to 11 in preorder.

We can now easily define the bijection between plane trees P and ballot sequences. Traverse P in preorder. Whenever we go down an edge (away from the root), record a 1. Whenever we go up an edge (toward the root), record a -1. For instance, for the plane tree P of Figure 1.8, the ballot sequence is (writing as usual $-$ for -1)

$$111 - 1 - - - 1 - 11 - 11 - 1 - - - .$$

It is also instructive to see directly how ballot sequences are related to the recurrence (1.1). Given a ballot sequence $\beta = a_1 a_2 \cdots a_{2n+2}$ of length $2(n+1)$,

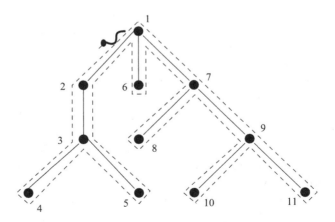

Figure 1.8. A plane tree traversed in preorder.

let k be the least nonnegative integer for which $a_1 + a_2 + \cdots + a_{2k+2} = 0$. Note that k exists and satisfies $0 \le k \le n$ since $a_1 + a_2 + \cdots + a_{2n+2} = 0$. We can then decompose β into the two ballot sequences $a_2 a_3 \cdots a_{2k+1}$ and $a_{2k+3} a_{2k+4} \cdots a_{2n+2}$ of lengths $2k$ and $2(n-k)$, respectively. Since the procedure is reversible, Equation (1.1) follows.

This decomposition of ballot sequences "demystifies" the bijection of de Bruijn-Morselt from plane trees to binary trees given above. Namely, binary trees B have an obvious decomposition (v, B_1, B_2) into the root v, the left subtree B_1 of v, and the right subtree B_2 of v that corresponds to the recurrence (1.1). On the other hand, we have just seen how to decompose (factor) a ballot sequence w as $1, u, -, v$ to yield the same recurrence. By iterating these decompositions we obtain a recursively defined bijection φ from binary trees to ballot sequences, namely, $\varphi(B) = 1, \varphi(B_1), -, \varphi(B_2)$, with the intial condition that the one-vertex tree corresponds to the ballot sequence $1-$. We also have given a simple bijection between plane trees and ballot sequences. Thus we have defined in a natural way a recursive bijection between plane trees and binary trees. This bijection is precisely the bijection of de Bruin and Morselt!

The above argument illustrates a general principle in describing bijections: suppose that $S_0, S_1, \ldots$ and $T_0, T_1, \ldots$ are two sequences of finite sets with $f(n) = \#S_n$ and $g(n) = \#T_n$. If we find combinatorial proofs that $f(n)$ and $g(n)$ satisfy the same recurrence, and if the initial conditions agree, then we have found a recursive bijection between S_n and T_n for all n. Once this bijection is found, we can look for a simpler, nonrecursive way to describe it.

(v)$\rightarrow$(ii) Given a parenthesization w of a string of $n+1$ x's, define a binary tree B_w recursively as follows. If $n = 0$, then $B_w = \emptyset$. Otherwise the outermost parentheses of w bracket a product $w = st$. Then B_w has a root vertex v with left subtree B_s and right subtree B_t. For instance, if $w = xx$, then B_w consists of a single vertex (the root). For the five binary trees of Figure 1.3, the corresponding parenthesizations are given by

$$((xx)x)x \quad (x(xx))x \quad x((xx)x) \quad x(x(xx)) \quad (xx)(xx).$$

We can also obtain a ballot sequence directly from a bracketing (with the outermost parentheses included) by converting each left parenthesis to a 1 and each x except the last into a -1, then deleting the right parentheses and the last x.

(vi)$\rightarrow$(iv) Given a Dyck path, replace an up step (a step $(1, 1)$) with a 1 and a down step (a step $(1, -1)$) with a -1, giving an obvious bijection with ballot sequences. $\qquad\square$

1.6. A Combinatorial Proof

The formula $C_n = \frac{1}{n+1}\binom{2n}{n}$ is so simple that we can ask whether there is a direct combinatorial proof. We will give one such proof in this subsection. One question that arises immediately is how to interpret the quantity $\frac{1}{n+1}\binom{2n}{n}$ combinatorially. Ideally we would like to find a bijection between some set C_n known to have C_n elements, in particular, the six sets of Theorem 1.5.1, and some set $\mathcal{D}_n$ that "obviously" contains $\frac{1}{n+1}\binom{2n}{n}$ elements. However, there is really no good choice for $\mathcal{D}_n$. What we actually need is to find an equivalence relation $\sim$ on some set $\mathcal{E}_n$ with $\binom{2n}{n}$ elements such that each equivalence class contains $n + 1$ elements. We can then give a bijection between C_n and the set $\mathcal{E}_n/\sim$ of equivalence classes of the equivalence relation $\sim$. Alternatively, we could do something analogous using the formula $C_n = \frac{1}{2n+1}\binom{2n+1}{n}$

Given a sequence $\alpha = a_1 a_2 \cdots a_k$, a *cyclic shift* or *conjugate* of α is a sequence

$$a_i a_{i+1} \cdots a_k a_1 a_2 \cdots a_{i-1}.$$

Thus α has k cyclic shifts, though they need not all be distinct. Also define a *strict ballot sequence* of length $2n + 1$ to be a sequence of $n + 1$ 1's and n -1's for which every nonempty partial sum is positive. Our proof follows easily from the following fundamental lemma (a special case of some much more general results).

1.6.1 Lemma. *Let $\alpha = a_1 a_2 \cdots a_{2n+1}$ be a sequence of $n + 1$ 1's and n -1's, where $n \geq 0$. Then all $2n + 1$ cyclic shifts of α are distinct, and exactly one of them is a strict ballot sequence.*

First proof. Since n and $n + 1$ are relatively prime, it follows easily that all cyclic shifts of α are distinct. The remainder of the proof is by induction on n. The statement is clearly true for $n = 0$ since the only sequence is 1. Assume for $n \geq 0$, and let $\alpha = a_1 a_2 \cdots a_{2n+3}$ be a sequence of $n + 2$ 1's and $n + 1$ -1's. For some j we must have $a_j = 1$ and $a_{j+1} = -1$, taking subscripts modulo $2n + 3$ if $j = 2n + 3$. Let $\beta = b_1 b_2 \cdots b_{2n+1}$ be the sequence obtained by removing a_j and a_{j+1} from α. By induction, β has a unique cyclic shift $\beta' = b_k b_{k+1} \cdots b_{k-1}$ which is a strict ballot sequence. Suppose that b_k corresponds to a_h. Let $\alpha' = a_h a_{h+1} \cdots a_{h-1}$. Now every nonempty partial sum of α' is positive, since α' agrees with β' until we reach a_j. Moreover, just before reaching a_j the partial sum is positive, then we add $a_j = 1$, then we add $a_{j+1} = -1$, and then the partial sums continue to agree with those of β'.

It remains to show that α' is unique. The cyclic shift beginning with a_j has the partial sum $a_j + a_{j+1} = 0$. The cyclic shift beginning with a_{j+1} has the partial sum $a_{j+1} = -1$. Any other cyclic shift of α' will begin with some b_i $(i \neq k)$ and have partial sums that include all partial sums of the cyclic shift $b_i b_{i+1} \cdots b_{i-1}$ of β. By induction, one of these partial sums must be negative, completing the proof.   □

Second proof (sketch). There is a nice geometric way to see that α has a unique cyclic shift α' which is a strict ballot sequence. We simply state the result without proof. Consider the word $\alpha\alpha = a_1 a_2 \cdots a_{2n+1} a_1 a_2 \cdots a_{2n+1}$. Represent this word as a lattice path P by starting at $(0,0)$ and then reading $\alpha\alpha$ from left to right, stepping one unit north if a term is equal to 1 and one unit east if a term is equal to -1. Let D be the lowest line parallel to $y = x$ that intersects P. Let v be the highest point of P on the line D. Then the first $2n + 1$ steps of the path beginning at v correspond to α'. Figure 1.9 shows the situation for $\alpha = 1-11-$ (writing as usual $-$ for -1). The first five steps beginning at v correspond to the cyclic shift $11-1-$.   □

Combinatorial proof. We can now give a combinatorial proof that $C_n = \frac{1}{n+1}\binom{2n}{n}$. First note that there is an obvious bijection between strict ballot sequences α of length $2n + 1$ and (ordinary) ballot sequences β of length $2n$, namely, remove the first 1 from α. Define two sequences of ± 1's for which the number of 1's is $n + 1$ and the number of -1's is n to be *equivalent* if they are cyclic shifts of each other. By Lemma 1.6.1 each equivalence class contains $2n + 1$ elements, of which exactly one is a strict ballot sequence. Since there are $\binom{2n+1}{n}$ sequences of $n + 1$ 1's and n -1's, we see that there are $\frac{1}{2n+1}\binom{2n+1}{n}$ equivalence classes, and hence the same number of strict ballot sequences, completing the proof.

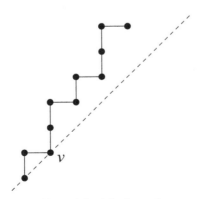

Figure 1.9. A lattice path.

Variant (suggested by J. Stembridge). Rather than dealing with *all* sequences of ± 1's for which 1 occurs $n + 1$ times and -1 occurs n times, we could consider only those sequences that begin with a 1. There are $\binom{2n}{n}$ such sequences. Each has $n + 1$ cyclic shifts beginning with a 1. Since a strict ballot sequence must begin with a 1, we get from Lemma 1.6.1 that exactly one of these $n + 1$ cyclic shifts is a strict ballot sequence. Hence the number of strict ballot sequences is $\frac{1}{n+1}\binom{2n}{n}$.

For alternative combinatorial proofs that $C_n = \frac{1}{n+1}\binom{2n}{n}$, see Problems A2–A3.

2

Bijective Exercises

We now come to the core of this monograph, the 214 combinatorial interpretations of Catalan numbers. We illustrate each item with the case $n = 3$, hoping that these illustrations will make any undefined terminology clear. Some of this terminology also appears in the Glossary. Solutions, hints, and references are given in the next section. In some instances two different items will agree as sets, but the descriptions of the sets will be different. Readers seeking to become experts on Catalan numbers are invited to take each pair (i_1, i_2) of distinct items and find a bijection (valid for all n) from the sets counted by i_1 to the sets counted by i_2, so $214 \cdot 213 = 45582$ bijections in all!

1. Triangulations of a convex $(n+2)$-gon into n triangles by $n-1$ diagonals that do not intersect in their interiors.

2. Total number of triangles with vertices $1, i, i+1$, $2 \le i \le n+1$, among all triangulations of a convex $(n+2)$-gon with vertices $1, 2, \ldots, n+2$ in clockwise order.

3. Binary parenthesizations or bracketings of a string of $n + 1$ letters.

$$(xx \cdot x)x \quad x(xx \cdot x) \quad (x \cdot xx)x \quad x(x \cdot xx) \quad xx \cdot xx$$

4. Binary trees with n vertices.

5. Complete binary trees (that is, binary trees such that every vertex has zero or two children) with $2n + 1$ vertices (or $n + 1$ endpoints).

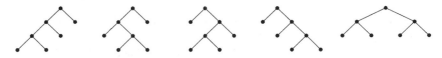

6. Plane trees with $n + 1$ vertices.

7. Planted (i.e., the root has one child) trivalent (i.e., all nonroot, nonleaf vertices have two children and hence three adjacent vertices) plane trees with $2n + 2$ vertices.

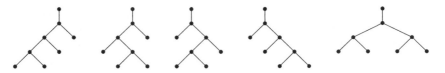

8. Plane trees with $n + 2$ vertices such that the rightmost path of each subtree of the root has even length.

9. Plane trees with $n - 1$ internal vertices (that is, vertices that are not endpoints), each having degree 1 or 2, such that vertices of degree 1 occur only on the rightmost path.

10. Plane trees with n vertices such that, going from left to right, all subtrees of the root first have an even number of vertices and then an odd number of vertices, with those subtrees with an odd number of vertices colored either red or blue.

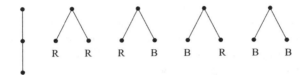

11. Plane trees with n vertices whose leaves at height one are colored red or blue.

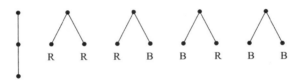

12. Plane trees with n internal vertices such that each vertex has at most two children and each left child of a vertex with two children is an internal vertex.

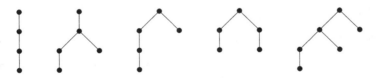

13. Plane trees for which every vertex has 0, 1, or 3 children, with a total of $n+1$ vertices with 0 or 1 child.

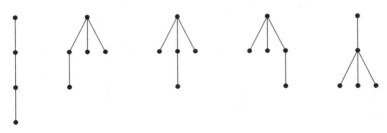

14. Bicolored (i.e., each vertex is colored black or white with no monochromatic edge) unrooted plane trees (i.e., unrooted trees with the subtrees at each vertex cyclically ordered) with $n+1$ vertices, rooted at an edge (marked ∗ below).

15. Plane trees with $n+1$ vertices, each nonroot vertex labeled by a positive integer, such that (a) leaves have label 1, (b) any nonroot, nonleaf vertex has a label no greater than the sum of its children's labels, and (c) the only edges with no right neighbor are those on the rightmost path from the root.

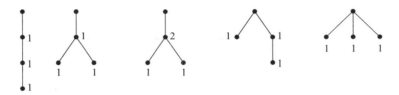

16. Increasing plane trees on the vertex set $[n+1]$ with increasing leaves in preorder (or from left to right), such that the path from the root 1 to $n+1$ contains every nonleaf vertex.

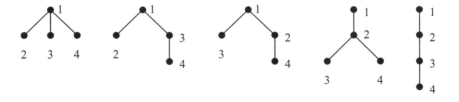

17. Total number of vertices v on the leftmost path (from the root) of a plane tree with $n > 0$ edges, such that v has exactly one child.

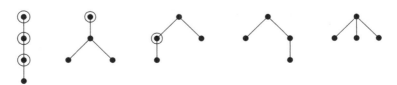

18. Plane trees with n vertices, where each single-child vertex v on the leftmost path (from the root) is colored either red or blue.

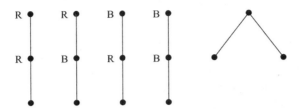

19. Rooted trees with n vertices such that each nonleaf vertex has either a left child, a middle child, a right child, or a left and right child, and such that every left child has a left child, a right child, or both.

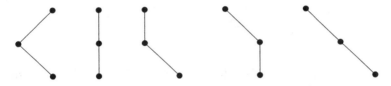

20. Induced subtrees with n edges, rooted at an endpoint, of the hexagonal lattice, up to translation and rotation.

21. Noncrossing increasing trees on the vertex set $[n+1]$, i.e., trees whose vertices are arranged in increasing order around a circle such that no edges cross in their interior, and such that all paths from vertex 1 are increasing.

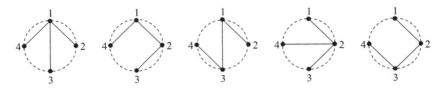

22. Nonnesting increasing trees on the vertex set $[n+1]$, i.e., trees with root 1 such that there do not exist vertices $h < i < j < k$ such that both hk and ij are edges, and such that every path from the root is increasing.

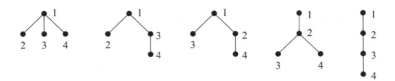

23. Vertices of height $n-1$ of the tree T defined by the property that the root has degree 2, and if the vertex x has degree k, then the children of x have degrees $2, 3, \ldots, k+1$.

24. Lattice paths from $(0,0)$ to (n,n) with steps $(0,1)$ or $(1,0)$, never rising above the line $y = x$.

25. Dyck paths of length $2n$, i.e., lattice paths from $(0,0)$ to $(2n,0)$ with steps $(1,1)$ and $(1,-1)$, never falling below the x-axis.

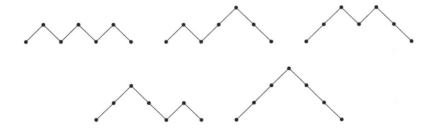

26. Dyck paths from $(0,0)$ to $(2n+2,0)$ such that any maximal sequence of consecutive steps $(1,-1)$ ending on the x-axis has odd length.

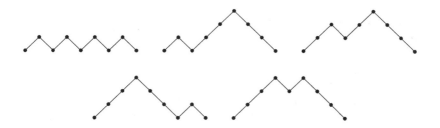

27. Dyck paths from $(0,0)$ to $(2n+2,0)$ with no peaks (an up step followed immediately by a down step) at height two.

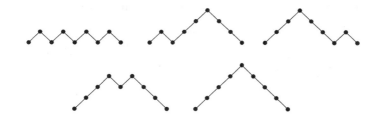

28. Left factors L of Dyck paths such that L has $n-1$ up steps.

29. Dyck paths of length $2n+2$ whose first down step is followed by another down step.

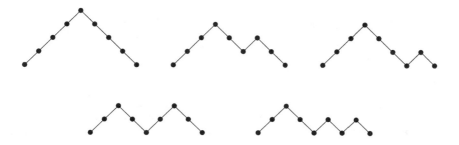

30. Dyck paths of length $2n+2$ with no peak at height 1 and having leftmost peak at height 2 or 3.

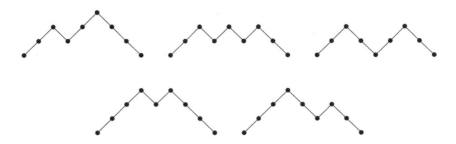

31. Dyck paths of length $2n + 2$ for which the terminal descent is of even
    length and all other descents (if any) to the x-axis are of odd length, where
    a *descent* is a maximal sequence of consecutive down steps (compare
    item 26).

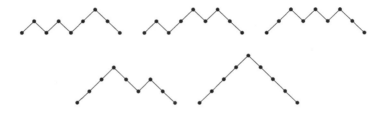

32. Dyck paths of length $4n$ such that every descent (maximal sequence of
    consecutive down steps) has length 2.

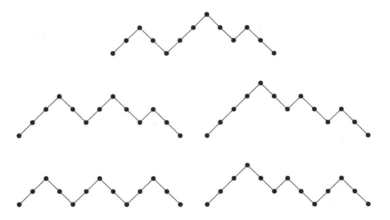

33. Dyck paths with $n - 1$ peaks and without three consecutive up steps or
    three consecutive down steps.

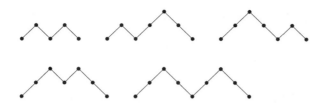

34. Dyck paths with n peaks such that there are no factors (consecutive steps) *UUU* and *UUDD*.

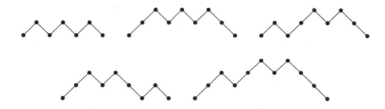

35. Dyck paths of length $2n+2$ whose second ascent (maximal sequence of consecutive up steps) has length 0 (i.e., there is only one ascent) or 2.

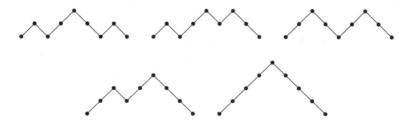

36. Dyck paths D from $(0,0)$ to $(2n+2,0)$ such that there is no horizontal line segment L with endpoints (i,j) and $(2n+2-i,j)$, with $i > 0$, such that the endpoints lie on P and no point of L lies above D (such line segments are called *centered tunnels*).

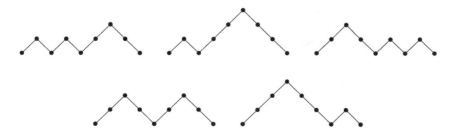

37. Points of the form $(m,0)$ on all Dyck paths from $(0,0)$ to $(2n-2,0)$.

38. Peaks of height one (or *hills*) in all Dyck paths from $(0,0)$ to $(2n,0)$.

39. Decompositions $AUBDC$ of Dyck paths of length $2n$ (regarded as a sequence of U's and D's), such that B is a Dyck path and such that A and C have the same length.

$$UUUDDD \quad UUUDDD \quad UUUDDD$$
$$UUDUDD \quad UDUDUD$$

40. Motzkin paths (as defined in Problem A49(d)) from $(0,0)$ to $(n-1,0)$ with the steps $(1,0)$ colored either red or blue.

41. Motzkin paths from $(0,0)$ to $(n,0)$ with the steps $(1,0)$ colored either red or blue, and with no red $(1,0)$ steps on the x-axis.

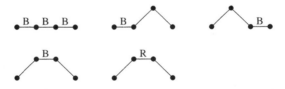

42. Peakless Motzkin paths having a total of n up and level steps.

43. Motzkin paths $a_1,\ldots,a_{2n-2}$ from $(0,0)$ to $(2n-2,0)$ such that each odd step a_{2i+1} is either $(1,0)$ (straight) or $(1,1)$ (up), and each even step a_{2i} is either $(1,0)$ (straight) or $(1,-1)$ (down).

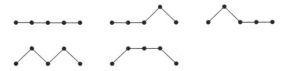

44. Motzkin paths with positive integer weights on the vertices of the path such that the sum of the weights is n.

45. Schröder paths as in Problem A50(g) from $(0,0)$ to $(2n,0)$ with no peaks, i.e., no up step followed immediately by a down step.

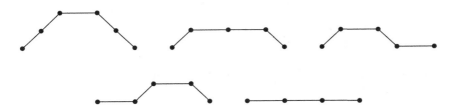

46. Schröder paths from $(0,0)$ to $(2(n-1),0)$ with peaks allowed only at the x-axis.

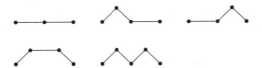

47. Schröder paths from $(0,0)$ to $(2n,0)$ with neither peaks nor level steps at odd height.

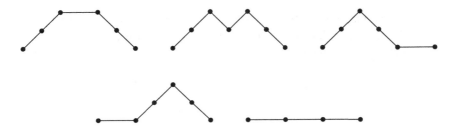

48. Schröder paths from $(0,0)$ to $(2n-2,0)$ with no valleys, i.e., no down step followed immediately by an up step.

49. Schröder paths from $(0,0)$ to $(2n-2,0)$ with no double rises, i.e., no two
    consecutive up steps (for $n = 3$ the set of paths is coincidentally the same
    as for item 46).

50. Schröder paths from $(0,0)$ to $(2n,0)$ with no level steps on the x-axis and
    no double rises.

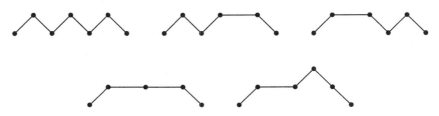

51. Paths on $\mathbb{N}$ of length $n-1$ from 0 to 0 with steps 1, -1, 0, 0* , i.e., there
    are two ways to take a step by standing still (the steps in the path are
    given below).

$$0,0 \qquad 0^*,0 \qquad 0,0^* \qquad 0^*,0^* \qquad 1,-1$$

52. Lattice paths from $(0,0)$ to $(n-1,n-1)$ with steps $(0,1)$, $(1,0)$, and
    $(1,1)$, never going below the line $y = x$, such that the steps $(1,1)$ only
    appear on the line $y = x$.

53. Lattice paths from $(0,0)$ with $n-1$ up steps $(1,1)$ and $n-1$ down steps
    $(1,-1)$, with no peaks at height $h \leq 0$.

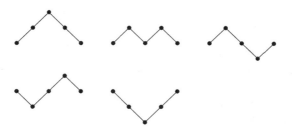

54. Lattice paths from $(0,0)$ with n up steps $(1,1)$ and n down steps $(1,-1)$, with no peaks at height $h \leq 1$.

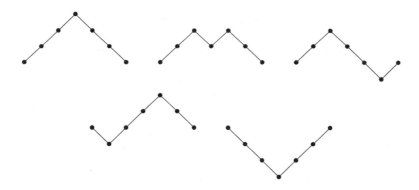

55. Lattice paths from $(0,0)$ with $n+1$ steps $(1,1)$ and $n-1$ steps $(1,-1)$, such that the interior vertices with even x-coordinate lie strictly above the line joining the initial and terminal points.

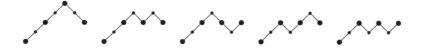

56. Lattice paths of length $n-1$ from $(0,0)$ to the x-axis with steps $(\pm 1, 0)$ and $(0, \pm 1)$, never going below the x-axis.

$$(-1,0) + (-1,0) \qquad (-1,0) + (1,0) \qquad (0,1) + (0,-1)$$

$$(1,0) + (-1,0) \qquad (1,0) + (1,0)$$

57. (Unordered) pairs of lattice paths with $n+1$ steps each, starting at $(0,0)$, using steps $(1,0)$ or $(0,1)$, ending at the same point, and only intersecting at the beginning and end.

58. (Unordered) pairs of lattice paths with $n - 1$ steps each, starting at $(0,0)$, using steps $(1,0)$ or $(0,1)$, ending at the same point, such that one path never rises above the other path.

59. n nonintersecting chords joining $2n$ points on the circumference of a circle.

60. Joining some of the vertices of a convex $(n - 1)$-gon by disjoint line segments, and circling a subset of the remaining vertices.

61. *Noncrossing (complete) matchings* on $2n$ vertices, i.e., ways of connecting $2n$ points in the plane lying on a horizontal line by n nonintersecting arcs, each arc connecting two of the points and lying above the points.

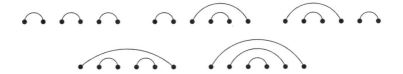

62. Ways of drawing in the plane $n + 1$ points lying on a horizontal line L and n arcs connecting them such that (a) the arcs do not pass below L, (b) the graph thus formed is a tree, (c) no two arcs intersect in their interiors

(i.e., the arcs are noncrossing), and (d) at every vertex, all the arcs exit in the same direction (left or right).

63. Ways of drawing in the plane $n+1$ points lying on a horizontal line L and n arcs connecting them such that (a) the arcs do not pass below L, (b) the graph thus formed is a tree, (c) no arc (including its endpoints) lies strictly below another arc, and (d) at every vertex, all the arcs exit in the same direction (left or right).

64. *Nonnesting matchings* on $[2n]$, i.e., ways of connecting $2n$ points in the plane lying on a horizontal line by n arcs, each arc connecting two of the points and lying above the points, such that no arc is contained entirely below another.

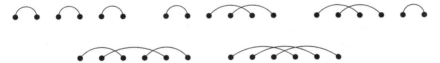

65. Ways of connecting $2n$ points in the plane lying on a horizontal line by n vertex-disjoint arcs, each arc connecting two of the points and lying above the points, such that the following condition holds: for every edge e let $n(e)$ be the number of edges e' that nest e (i.e., e lies below e'), and let $c(e)$ be the number of edges e' that begin to the left of e and that cross e. Then $n(e) - c(e) = 0$ or 1.

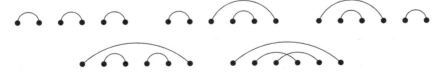

66. Ways of connecting any number of points in the plane lying on a horizontal line by nonintersecting arcs lying above the points, such that the total number of arcs and isolated points is $n-1$ and no isolated point lies below an arc.

67. Ways of connecting n points in the plane lying on a horizontal line by noncrossing arcs above the line such that if two arcs share an endpoint p, then p is a left endpoint of both the arcs.

68. Ways of connecting $n+1$ points in the plane lying on a horizontal line by noncrossing arcs above the line such that no arc connects adjacent points and the right endpoints of the arcs are all distinct.

69. Ways of connecting $n+1$ points in the plane lying on a horizontal line by n noncrossing arcs above the line such that the left endpoints of the arcs are distinct.

70. Ways of connecting $2n-2$ points labeled $1,2,\dots,2n-2$ lying on a horizontal line by nonintersecting arcs above the line such that the left endpoint of each arc is odd and the right endpoint is even.

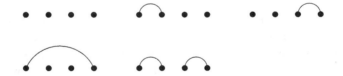

71. Noncrossing matchings of some vertex set $[k]$ into n components such that no two consecutive vertices form an edge.

72. Lattice paths in the first quadrant with n steps from $(0,0)$ to $(0,0)$, where each step is of the form $(\pm 1, \pm 1)$, or goes from $(2k,0)$ to $(2k,0)$ or $(2(k+1),0)$, or goes from $(0,2k)$ to $(0,2k)$ or $(0,2(k+1))$.

$$(0,0) \to (0,0) \to (0,0) \to (0,0)$$

$$(0,0) \to (0,0) \to (1,1) \to (0,0)$$

$$(0,0) \to (1,1) \to (0,0) \to (0,0)$$

$$(0,0) \to (2,0) \to (1,1) \to (0,0)$$

$$(0,0) \to (0,2) \to (1,1) \to (0,0)$$

73. Lattice paths from $(0,0)$ to $(n,-n)$ such that (a) from a point (x,y) with $x < 2y$ the allowed steps are $(1,0)$ and $(0,1)$, (b) from a point (x,y) with $x > 2y$ the allowed steps are $(0,-1)$ and $(1,-1)$, (c) from a point $(2y,y)$ the allowed steps are $(0,1)$, $(0,-1)$, and $(1,-1)$, and (d) it is forbidden to enter a point $(2y+1,y)$.

74. Symmetric parallelogram polyominoes (where parallelogram polyomino is defined in the solution to item 57) of perimeter $4(2n+1)$ such that the horizontal boundary steps on each level (equivalently, vertical boundary steps with fixed x-coordinate) form an unbroken line.

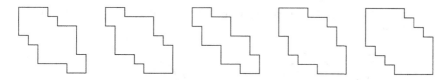

75. All horizontal chords in the nonintersecting chord diagrams of item 59 (with the vertices drawn so that one of the diagrams has n horizontal chords).

76. *Kepler towers* with n *bricks*, i.e., sets of concentric circles, with "bricks" (arcs) placed on each circle, as follows: the circles come in sets called *walls* from the center outward. The circles (or *rings*) of the ith wall are divided into 2^i equal arcs, numbered $1,2,\ldots,2^i$ clockwise from due north. Each brick covers an arc and extends slightly beyond the endpoints of the arc. No two consecutive arcs can be covered by bricks. The first (innermost) arc within each wall has bricks at positions $1,3,5,\ldots,2^i-1$. Within each wall, each brick B not on the innermost ring must be

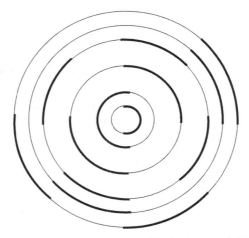

Figure 2.1. A Kepler tower with three walls, six rings, and thirteen bricks.

supported by another brick B' on the next ring toward the center, i.e., some ray from the center must intersect both B and B'. Finally, if $i > 1$ and the ith wall is nonempty, then wall $i - 1$ must also be nonempty. Figure 2.1 shows a Kepler tower with three walls, six rings, and thirteen bricks.

77. Ballot sequences, i.e., sequences of n 1's and n -1's such that every partial sum is nonnegative (with -1 denoted simply as $-$ below).

     111——   11–1—   11——1–   1–11—   1–1–1–

78. Sequences $1 \le a_1 \le \cdots \le a_n$ of integers with $a_i \le i$.

           111   112   113   122   123

79. Sequences $a_1 < a_2 < \cdots < a_{n-1}$ of integers satisfying $1 \le a_i \le 2i$.

           12   13   14   23   24

80. Sequences $a_1, a_2, \ldots, a_n$ of integers such that $a_1 = 0$ and $0 \le a_{i+1} \le a_i + 1$.

       000   001   010   011   012

81. Sequences $a_1, a_2, \ldots, a_{n-1}$ of integers such that $a_i \leq 1$ and all partial sums are nonnegative.

$$0,0 \quad 0,1 \quad 1,-1 \quad 1,0 \quad 1,1$$

82. Sequences $a_1, a_2, \ldots, a_n$ of integers such that $a_i \geq -1$, all partial sums are nonnegative, and $a_1 + a_2 + \cdots + a_n = 0$.

$$0,0,0 \quad 0,1,-1 \quad 1,0,-1 \quad 1,-1,0 \quad 2,-1,-1$$

83. Sequences $a_1, a_2, \ldots, a_n$ of integers such that $0 \leq a_i \leq n - i$, and such that if $i < j$, $a_i > 0$, $a_j > 0$, and $a_{i+1} = a_{i+2} = \cdots = a_{j-1} = 0$, then $j - i > a_i - a_j$.

$$000 \quad 010 \quad 100 \quad 200 \quad 110$$

84. Sequences $a_1, a_2, \ldots, a_n$ of integers such that $i \leq a_i \leq n$ and such that if $i \leq j \leq a_i$, then $a_j \leq a_i$.

$$123 \quad 133 \quad 223 \quad 323 \quad 333$$

85. Sequences $a_1, a_2, \ldots, a_n$ of integers such that $1 \leq a_i \leq i$ and such that if $a_i = j$, then $a_{i-r} \leq j - r$ for $1 \leq r \leq j - 1$.

$$111 \quad 112 \quad 113 \quad 121 \quad 123$$

86. Sequences $(a_1, \ldots, a_n)$ of nonnegative integers satisfying $a_1 + \cdots + a_i \geq i$ and $\sum a_j = n$.

$$111 \quad 120 \quad 210 \quad 201 \quad 300$$

87. Sequences $1 \leq a_1 \leq a_2 \leq \cdots \leq a_n \leq n$ of integers with exactly one fixed point, i.e., exactly one value of i for which $a_i = i$.

$$111 \quad 112 \quad 222 \quad 233 \quad 333$$

88. Sequences $a_1, a_2, \ldots, a_n$ of positive integers such that the first appearance of $i \geq 1$ occurs before the first appearance of $i + 1$, and there are no subsequences of the form $abab$.

$$111 \quad 112 \quad 121 \quad 122 \quad 123$$

89. Sequences $1 \leq a_1 < a_2 < \cdots < a_n \leq 2n$ such that $a_i \geq 2i$.

$$246 \quad 256 \quad 346 \quad 356 \quad 456$$

90. Sequences $1 \leq a_1 < a_2 < \cdots < a_n \leq 2n$ such that $a_i \leq 2i - 1$.

$$123 \quad 124 \quad 125 \quad 134 \quad 135$$

91. Number of pairs of integer sequences $1 \le a_1 < a_2 < \cdots < a_k \le n$ and $1 \le b_1 < b_2 < \cdots < b_k \le n$ for which there exists a permutation $w \in \mathfrak{S}_n$ whose left-to-right maxima (or records) are given by $w(a_i) = b_i$.

$$(123, 123) \quad (12, 13) \quad (13, 23) \quad (12, 23) \quad (1, 3)$$

92. n-tuples $(a_1, a_2, \ldots, a_n)$ of integers $a_i \ge 2$ such that in the sequence $1 a_1 a_2 \cdots a_n 1$, each a_i divides the sum of its two neighbors.

$$14321 \quad 13521 \quad 13231 \quad 12531 \quad 12341$$

93. Sequences $a_0, \ldots, a_n$ defined recursively as follows: 1 is an allowed sequence (the case $n = 0$). A sequence $a_0, \ldots, a_n$ is obtained from one for $n - 1$ either by appending 1 at the end, or by inserting between two consecutive terms their sum.

$$1111 \quad 1211 \quad 1121 \quad 1231 \quad 1321$$

94. Sequences $a_1, \ldots, a_{2n}$ of nonnegative integers with $a_1 = 1$, $a_{2n} = 0$ and $a_i - a_{i-1} = \pm 1$.

$$123210 \quad 121210 \quad 121010 \quad 101210 \quad 101010$$

95. Sequences $w = a_1 \cdots a_{n+1}$ of nonnegative integers such that $a_{k+1} \ge a_k - 1$ for all $1 \le k \le n$, and such that if w contains an $i \ge 1$, then the first i lies between two $i - 1$'s (not necessarily consecutively).

$$0000 \quad 0010 \quad 0100 \quad 0101 \quad 0110$$

96. Sequences $a_1, \ldots, a_n$ of nonnegative integers with $a_1 = 0$ and for $2 \le i \le n$,

$$\max\{a_1, \ldots, a_{i-1}\} - 1 \le a_i \le \mathrm{asc}(a_1, \ldots, a_{i-1}) + 1,$$

where $\mathrm{asc}(a_1, \ldots, a_{i-1})$ denotes the number of ascents (indices $1 \le j \le n - 1$ for which $a_j < a_{j+1}$) of the sequence $(a_1, \ldots, a_{i-1})$.

$$000 \quad 001 \quad 010 \quad 011 \quad 012$$

97. Sequences $a_1, \ldots, a_{n+1}$ of nonnegative integers such that $\sum_{i=1}^{n+1} 2^{-a_i} = 1$ and for all $2 \le j \le n+1$ the number $2^{a_j} \sum_{i=1}^{j-1} 2^{-a_i}$ is an integer.

$$1233 \quad 1332 \quad 2222 \quad 2331 \quad 3321$$

98. Sequences of $n - 1$ 1's and any number of -1's such that every partial sum is nonnegative.

$$1,1 \quad 1,1,-1 \quad 1,-1,1 \quad 1,1,-1,-1 \quad 1,-1,1,-1$$

99. Sequences $a_1, a_2, \ldots, a_n$ of nonnegative integers satisfying: (a) if k is the smallest integer with $a_k \neq 0$ then $a_k = 1$; (b) if k is the largest integer with $a_k \neq 0$ then $a_k = 1$ or $a_k = 2$; (c) if a_m is even and $a_{m+1} > a_m$ then $a_{m+1} = a_m + 1$; (d) if a_m is odd and $a_{m+1} > a_m$ then $a_{m+1} = a_m + 1$ or $a_{m+1} = a_m + 2$; (e) if a_m is even and $a_{m+1} < a_m$ then $a_{m+1} = a + m - 1$ or $a_{m+1} = a + m - 2$; and (f) if a_m is odd and $a_{m+1} < a_m$ then $a_{m+1} = a_m - 1$.

$$00 \qquad 10 \qquad 01 \qquad 11 \qquad 12$$

100. Subsequences of the sequence

$$(a_1, a_2, \ldots, a_{2n-2}) = (1, -1, 1, -1, \ldots, 1, -1)$$

that are ballot sequences (as defined in Section 1.5).

$$\emptyset \qquad (a_1, a_2) \qquad (a_1, a_4) \qquad (a_3, a_4) \qquad (a_1, a_2, a_3, a_4)$$

101. Sequences $a_1, a_2, \ldots, a_{2n-2}$ of $n-1$ 1's and $n-1$ -1's such that if $a_i = -1$ then either $a_{i+1} = a_{i+2} = \cdots = a_{2n-2} = -1$ or $a_{i+1} + a_{i+2} + \cdots + a_{i+j} > 0$ for some $j \geq 1$.

$$1,1,-1,-1 \qquad 1,-1,1,-1 \qquad -1,1,1,-1 \qquad -1,1,-1,1 \qquad -1,-1,1,1$$

102. Sequences $a_1, a_2, \ldots, a_n$ of integers such that $a_1 = 1$, $a_n = \pm 1$, $a_i \neq 0$, and $a_{i+1} \in \{a_i, a_i + 1, a_i - 1, -a_i\}$ for $2 \leq i \leq n$.

$$1,1,1 \qquad 1,1,-1 \qquad 1,-1,1 \qquad 1,-1,-1 \qquad 1,2,1$$

103. Sequences $a_1, a_2, \ldots, a_n$ of nonnegative integers such that $a_j = \#\{i : i < j, a_i < a_j\}$ for $1 \leq j \leq n$.

$$000 \qquad 002 \qquad 010 \qquad 011 \qquad 012$$

104. Sequences $a_1, a_2, \ldots, a_{n-1}$ of nonnegative integers such that each nonzero term initiates a factor (subsequence of consecutive elements) whose length is equal to its sum.

$$00 \quad 01 \quad 10 \quad 11 \quad 20$$

105. Sequences $a_1, a_2, \ldots, a_{2n+1}$ of positive integers such that $a_{2n+1} = 1$, some $a_i = n+1$, the first appearance of $i+1$ follows the first appearance of i, no two consecutive terms are equal, no pair ij of integers occur more than once as a factor (i.e., as two consecutive terms), and if ij is a factor then so is ji.

$$1213141 \qquad 1213431 \qquad 1232141 \qquad 1232421 \qquad 1234321$$

106. Sequences $a_1, \ldots, a_n$ of nonnegative integers for which there exists a distributive lattice of rank n with a_i join-irreducibles of rank i, $1 \leq i \leq n$.

$$300 \quad 210 \quad 120 \quad 201 \quad 111$$

107. Pairs of sequences $1 \leq i_1 < i_2 < \cdots < i_k \leq n-1$ and $2 \leq j_1 < \cdots < j_k \leq n$ such that $i_r < j_r$ for all r.

$$(\emptyset, \emptyset) \quad (1,2) \quad (1,3) \quad (2,3) \quad (12,23)$$

108. Compositions of n whose parts equal to k are colored with one of C_{k-1} colors (colors are indicated by subscripts below).

$$1+1+1 \quad 1+2 \quad 2+1 \quad 3_a \quad 3_b$$

109. Pairs (α, β) of compositions of n with the same number of parts, such that $\alpha \geq \beta$ (dominance order, i.e., $\alpha_1 + \cdots + \alpha_i \geq \beta_1 + \cdots + \beta_i$ for all i).

$$(111,111) \quad (12,12) \quad (21,21) \quad (21,12) \quad (3,3)$$

110. Equivalence classes B of words in the alphabet $[n-1]$ under the equivalence relation $uijv \sim ujiv$ for any words u, v (possibly empty) and any $i, j \in [n-1]$ satisfying $|i-j| \geq 2$, where no equivalence class can contain a word with three consecutive letters that are not distinct.

$$\{\emptyset\} \quad \{1\} \quad \{2\} \quad \{12\} \quad \{21\}$$

(For $n = 4$ a representative of each class is given by $\emptyset$, 1, 2, 3, 12, 21, 13, 23, 32, 123, 132, 213, 321, 2132. The equivalence class containing 1321, for instance, is not counted because it also contains 3121.)

111. Partitions $\lambda = (\lambda_1, \ldots, \lambda_{n-1})$ with $\lambda_1 \leq n-1$ (so the diagram of λ is contained in an $(n-1) \times (n-1)$ square), such that if $\lambda' = (\lambda_1', \lambda_2', \ldots)$ denotes the conjugate partition to λ, then $\lambda_i' \geq \lambda_i$ whenever $\lambda_i \geq i$.

$$(0,0) \quad (1,0) \quad (1,1) \quad (2,1) \quad (2,2)$$

112. Integer partitions that are both n-cores and $(n+1)$-cores, in the terminology of [65, Exercise 7.59(d)].

$$\emptyset \quad 1 \quad 2 \quad 11 \quad 311$$

113. Permutations $a_1 a_2 \cdots a_{2n}$ of the multiset $\{1^2, 2^2, \ldots, n^2\}$ such that: (a) the first occurrences of $1, 2, \ldots, n$ appear in increasing order, and (b) there is no subsequence of the form $\alpha\beta\alpha\beta$.

$$112233 \quad 112332 \quad 122331 \quad 123321 \quad 122133$$

114. Permutations $a_1a_2 \cdots a_{2n}$ of the set $[2n]$ such that: (a) $1,3,\ldots,2n-1$ appear in increasing order, (b) $2,4,\ldots,2n$ appear in increasing order, and (c) $2i-1$ appears before $2i$, $1 \leq i \leq n$.

$$123456 \quad 123546 \quad 132456 \quad 132546 \quad 135246$$

115. Permutations $a_1a_2 \cdots a_n$ of $[n]$ with longest decreasing subsequence of length at most two (i.e., there does not exist $i < j < k$, $a_i > a_j > a_k$), called 321-*avoiding* permutations.

$$123 \quad 213 \quad 132 \quad 312 \quad 231$$

116. Permutations $a_1a_2 \cdots a_n$ of $[n]$ for which there does not exist $i < j < k$ and $a_j < a_k < a_i$ (called 312-*avoiding* permutations).

$$123 \quad 132 \quad 213 \quad 231 \quad 321$$

117. Permutations w of $[2n]$ with n cycles of length two, such that the product $(1,2,\ldots,2n) \cdot w$ has $n+1$ cycles

$$(1,2,3,4,5,6)(1,2)(3,4)(5,6) = (1)(2,4,6)(3)(5)$$
$$(1,2,3,4,5,6)(1,2)(3,6)(4,5) = (1)(2,6)(3,5)(4)$$
$$(1,2,3,4,5,6)(1,4)(2,3)(5,6) = (1,3)(2)(4,6)(5)$$
$$(1,2,3,4,5,6)(1,6)(2,3)(4,5) = (1,3,5)(2)(4)(6)$$
$$(1,2,3,4,5,6)(1,6)(2,5)(3,4) = (1,5)(2,4)(3)(6)$$

118. Permutations u of $[n]$ such that u and $u(1,2,\ldots,n)$ have a total of $n+1$ cycles, the largest possible (the permutations u are called permutations of *genus* 0).

$$(1)(2)(3) \quad (1,2)(3) \quad (1,3)(2) \quad (1)(2,3) \quad (1,3,2)$$

119. Permutations $a_1a_2 \cdots a_n$ of $[n]$ that can be put in increasing order on a single stack, defined recursively as follows: if $\emptyset$ is the empty sequence, then let $S(\emptyset) = \emptyset$. If $w = unv$ is a sequence of distinct integers with largest term n, then $S(w) = S(u)S(v)n$. A *stack-sortable* permutation w is one for which $S(w) = w$.

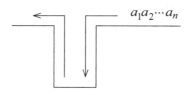

For example,

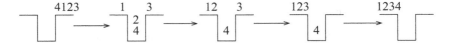

$$123 \quad 132 \quad 213 \quad 312 \quad 321$$

120. Permutations $a_1 a_2 \cdots a_n$ of $[n]$ that can be put in increasing order on two parallel queues. Now the picture is

$$123 \quad 132 \quad 213 \quad 231 \quad 312$$

121. Fixed-point free involutions w of $[2n]$ such that if $i < j < k < l$ and $w(i) = k$, then $w(j) \neq l$ (in other words, 3412-avoiding fixed-point free involutions).

$$(12)(34)(56) \quad (14)(23)(56) \quad (12)(36)(45) \quad (16)(23)(45) \quad (16)(25)(34)$$

122. Cycles of length $2n+1$ in $\mathfrak{S}_{2n+1}$ with descent set $\{n\}$.

$$2371456 \quad 2571346 \quad 3471256 \quad 3671245 \quad 5671234$$

123. Baxter permutations (as defined in [65, Exercise 6.55]) of $[2n]$ or of $[2n+1]$ that are reverse alternating (as defined at the end of Section 3.16) and whose inverses are reverse alternating

$$132546 \quad 153426 \quad 354612 \quad 561324 \quad 563412$$

$$1325476 \quad 1327564 \quad 1534276 \quad 1735462 \quad 1756342$$

124. Permutations w of $[n]$ such that if w has ℓ inversions then for all pairs of sequences $(a_1, a_2, \ldots, a_\ell), (b_1, b_2, \ldots, b_\ell) \in [n-1]^\ell$ satisfying

$$w = s_{a_1} s_{a_2} \cdots s_{a_\ell} = s_{b_1} s_{b_2} \cdots s_{b_\ell},$$

where s_j is the adjacent transposition $(j, j+1)$, we have that the ℓ-element *multisets* $\{a_1, a_2, \ldots, a_\ell\}$ and $\{b_1, b_2, \ldots, b_\ell\}$ are equal (thus, for example, $w = 321$ is not counted, since $w = s_1 s_2 s_1 = s_2 s_1 s_2$, and the multisets $\{1, 2, 1\}$ and $\{2, 1, 2\}$ are not equal).

$$123 \quad 132 \quad 213 \quad 231 \quad 312$$

125. Permutations w of $[n]$ with the following property: Suppose that w has ℓ inversions, and let

$$R(w) = \{(a_1,\ldots,a_\ell) \in [n-1]^\ell : w = s_{a_1} s_{a_2} \cdots s_{a_\ell}\},$$

where s_j is as in item 124. Then

$$\sum_{(a_1,\ldots,a_\ell) \in R(w)} a_1 a_2 \cdots a_\ell = \ell!.$$

$$R(123) = \{\emptyset\}, \quad R(213) = \{(1)\}, \quad R(231) = \{(1,2)\}$$

$$R(312) = \{(2,1)\}, \quad R(321) = \{(1,2,1),(2,1,2)\}$$

126. Indecomposable (as defined in [64, Exercise 1.128]) w-avoiding permutations of $[n+1]$, where w is any of 321, 312, or 231.

$$w = 321: \quad 4123 \quad 3142 \quad 2413 \quad 3412 \quad 2341$$

$$w = 312: \quad 4321 \quad 3421 \quad 2431 \quad 2341 \quad 3241$$

$$w = 231: \quad 4123 \quad 4132 \quad 4213 \quad 4312 \quad 4321$$

127. Decomposable 213-avoiding permutations of $[n+1]$, where w is either of 213 or 132.

$$w = 213: \quad 1432 \quad 1423 \quad 1342 \quad 1243 \quad 1234$$

$$w = 132: \quad 1234 \quad 2134 \quad 2314 \quad 3124 \quad 3214$$

128. Weak ordered partitions (P,V,A,D) of $[n]$ into four blocks such that there exists a permutation $w = a_1 a_2 \cdots a_n \in \mathfrak{S}_n$ (with $a_0 = a_{n+1} = 0$) satisfying

$$P = \{i \in [n] : a_{i-1} < a_i > a_{i+1}\}$$

$$V = \{i \in [n] : a_{i-1} > a_i < a_{i+1}\}$$

$$A = \{i \in [n] : a_{i-1} < a_i < a_{i+1}\}$$

$$D = \{i \in [n] : a_{i-1} > a_i > a_{i+1}\}.$$

$$(3,\emptyset,12,\emptyset) \quad (3,\emptyset,1,2) \quad (23,1,\emptyset,\emptyset) \quad (3,\emptyset,2,1) \quad (3,\emptyset,\emptyset,12)$$

129. Factorizations $(1,2,\ldots,n+1) = (a_1,b_1)(a_2,b_2)\cdots(a_n,b_n)$ of the cycle $(1,2,\ldots,n+1)$ into n transpositions (a_i,b_i) such that $a_i < b_i$ for all i and $a_1 \le a_2 \le \cdots \le a_n$ (where we multiply permutations right to left).

$$(14)(13)(12) \quad (14)(12)(23) \quad (13)(12)(34)$$

$$(12)(24)(23) \quad (12)(23)(34)$$

130. Sequences $a_1, a_2, \ldots, a_p$ such that (a) if $u_1, \ldots, u_{n-1}$ is any fixed ordering (chosen at the beginning) of the adjacent transpositions $s_i = (i, i+1)$ $\in \mathfrak{S}_n$, then $u_{a_1} u_{a_2} \cdots u_{a_p}$ is reduced, i.e., has p inversions as a permutation in $\mathfrak{S}_n$, and (b) if the descent set of the sequence $a_1 a_2 \cdots a_p$ is $1 \leq j_1 < \cdots < j_k \leq n-1$, then

$$\{a_1, \ldots, a_{j_1}\} \supset \{a_{j_1+1}, \ldots, a_{j_2}\} \supset \cdots \supset \{a_{j_k+1}, \ldots, a_p\}.$$

For instance, if $n = 10$ and $u_i = s_i$ for all i, then an example of a sequence being counted is 1234567891246781 67. For $n = 3$ and $u_i = s_i$, we get the sequences

$$\emptyset \quad 1 \quad 2 \quad 12 \quad 121$$

131. Shuffles of the permutation $12 \cdots n$ with itself, i.e., permutations of the multiset $\{1^2, 2^2, \ldots, n^2\}$ which are a union of two disjoint subsequences $12 \cdots n$ (equivalently, there is no weakly decreasing subsequence of length three).

$$112233 \quad 112323 \quad 121233 \quad 121323 \quad 123123$$

132. *Unique* shuffles of the permutation $12 \cdots (n+1)$ with itself, i.e., permutations of the multiset $\{1^2, 2^2, \ldots, (n+1)^2\}$ which are a union of two disjoint subsequences $12 \cdots (n+1)$ in exactly one way

$$12132434 \quad 12134234 \quad 12312434 \quad 12314234 \quad 12341234$$

133. Pats $w \in \mathfrak{S}_{n+1}$, defined recursively as follows: a word a in the alphabet $\mathbb{P}$ is a *pat* if a is a singleton or can be uniquely factored $a = bc$ such that every letter of b is greater than every letter of c, and such that the reverse of a and b are pats.

$$2431 \quad 3241 \quad 3412 \quad 4132 \quad 4213$$

134. Permutations of the multiset $\{1^2, 2^2, \ldots, (n+1)^2\}$ such that the first appearances of $1, 2, \ldots, n, n+1$ occur in that order, and between the two appearances of i there is exactly one $i+1$ for $i = 1, 2, \ldots, n$.

$$12132434 \quad 12132434 \quad 12314234 \quad 12314234 \quad 12341234$$

135. Permutations $a_1 \cdots a_n \in \mathfrak{S}_n$ for which there does not exist $1 \leq i < j \leq n$ satisfying either of the conditions $i < j \leq a_i < a_j$ or $a_i < a_j < i < j$.

$$123 \quad 132 \quad 213 \quad 312 \quad 321$$

136. Permutations $w \in \mathfrak{S}_{n+1}$ of genus 0 (as defined in item 118) satisfying $w(1) = 1$.

$$1234 \quad 1243 \quad 1324 \quad 1342 \quad 1432$$

137. Permutations $w \in \mathfrak{S}_{n+1}$ of genus 0 satisfying $w(1) = 2$.

$$2134 \quad 2143 \quad 2314 \quad 2341 \quad 2431$$

138. Permutations $w \in \mathfrak{S}_{n+2}$ of genus 0 satisfying $w(1) = 3$.

$$32145 \quad 32154 \quad 32415 \quad 32451 \quad 32541$$

139. Permutations $w \in \mathfrak{S}_{n+2}$ of genus 0 satisfying $w(1) = n + 2$.

$$52341 \quad 52431 \quad 53241 \quad 53421 \quad 54321$$

140. Permutations $w \in \mathfrak{S}_n$ satisfying the following condition: let $w = R_{s+1} R_s \cdots R_1$ be the factorization of w into ascending runs (so $s = \operatorname{des}(w)$, the number of descents of w). Let m_k and M_k be the smallest and largest elements in the run R_k. Let n_k be the number of symbols in R_k for $1 \leq k \leq s+1$; otherwise set $n_k = 0$. Define $N_k = \sum_{i \leq k} n_i$ for all $k \in \mathbb{Z}$. Then $m_{s+1} > m_s > \cdots > m_1$ and $M_i \leq N_{i+1}$ for $1 \leq i \leq s+1$.

$$123 \quad\quad 213 \quad\quad 231 \quad\quad 312 \quad\quad 321$$

141. Permutations $w \in \mathfrak{S}_n$ satisfying, in the notation of item 140 above, $m_{s+1} > m_s > \cdots > m_1$ and $m_{i+1} > N_{i-1} + 1$ for $1 \leq i \leq s$.

$$123 \quad\quad 213 \quad\quad 231 \quad\quad 312 \quad\quad 321$$

142. Permutations $w = a_1 a_2 \cdots a_{2n+1} \in \mathfrak{S}_{2n+1}$ that are symmetric (i.e., $a_i + a_{2n+2-i} = 2n+2$) and 123-avoiding.

$$5764213 \quad 6574132 \quad 6754312 \quad 7564231 \quad 7654321$$

143. Total number of fixed points (written in boldface below) of all 321-avoiding permutations $w \in \mathfrak{S}_n$.

$$\mathbf{123} \quad \mathbf{1}32 \quad 3\mathbf{1}2 \quad 2\mathbf{13} \quad 231$$

144. Total number of fixed points (written in boldface below) of all 132-avoiding permutations $w \in \mathfrak{S}_n$.

$$\mathbf{123} \quad 2\mathbf{13} \quad 231 \quad 3\mathbf{1}2 \quad 3\mathbf{2}1$$

145. 321-avoiding permutations $w \in \mathfrak{S}_{2n+1}$ such that $i \in [2n]$ is an excedance of w (i.e., $w(i) > i$) if and only if $w(i) - 1$ is not an excedance of w (so that w has exactly n excedances).

$$4512736 \quad 3167245 \quad 3152746 \quad 4617235 \quad 5671234$$

146. 321-avoiding alternating permutations in $\mathfrak{S}_{2n}$ (for $n \geq 2$) or in $\mathfrak{S}_{2n-1}$.

$$214365 \quad 215364 \quad 314265 \quad 315264 \quad 415263$$

$$21435 \quad 21435 \quad 31425 \quad 31524 \quad 41523$$

147. 321-avoiding reverse alternating permutations in $\mathfrak{S}_{2n-2}$ ($n \geq 3$) or in $\mathfrak{S}_{2n-1}$.

$$1324 \quad 2314 \quad 1423 \quad 2413 \quad 3412$$

$$13254 \quad 23154 \quad 14253 \quad 24153 \quad 34152$$

148. 321-avoiding permutations of $[n+1]$ whose first (least) ascent is even.

$$2134 \quad 2143 \quad 3124 \quad 3142 \quad 4123$$

149. 132-avoiding alternating permutations in $\mathfrak{S}_{2n-1}$ or in $\mathfrak{S}_{2n}$.

$$32415 \quad 42315 \quad 43512 \quad 52314 \quad 53412$$

$$435261 \quad 534261 \quad 546231 \quad 634251 \quad 645231$$

150. 132-avoiding reverse alternating permutations in $\mathfrak{S}_{2n}$ or in $\mathfrak{S}_{2n+1}$.

$$342516 \quad 452316 \quad 453612 \quad 562314 \quad 563412$$

$$4536271 \quad 5634271 \quad 5647231 \quad 6734251 \quad 6745231$$

151. 321-avoiding fixed-point-free involutions of $[2n]$.

$$214365 \quad 215634 \quad 341265 \quad 351624 \quad 456123$$

152. 321-avoiding involutions of $[2n-1]$ with one fixed point.

$$13254 \quad 14523 \quad 21354 \quad 21435 \quad 34125$$

153. 213-avoiding fixed-point-free involutions of $[2n]$.

$$456123 \quad 465132 \quad 564312 \quad 645231 \quad 654321$$

154. 213-avoiding involutions of $[2n-1]$ with one fixed point.

$$14523 \quad 15432 \quad 45312 \quad 52431 \quad 54321$$

155. 3412-avoiding (or noncrossing) involutions of a subset of $[n-1]$.

$$\emptyset \quad 1 \quad 2 \quad 12 \quad 21$$

156. Distinct sets S obtained by starting with a permutation $w \in \mathfrak{S}_{2n}$ and continually crossing out the smallest element and then removing the leftmost element and placing it in S, until no elements remain. For

instance, if $w = 324165$, then cross out 1 and place 3 in S, then cross out 2 and place 4 in S, and then cross out 5 and place 6 in S, obtaining $S = \{3,4,6\}$.

$$\{2,4,6\} \quad \{2,5,6\} \quad \{3,4,6\} \quad \{3,5,6\} \quad \{4,5,6\}$$

157. Ways two persons can each start with 0 and alternatingly add positive integers to their numbers so that they first have equal numbers when that number is n (notation such as $1,2;4,3;5,5$ means that the first person adds 1 to 0 to obtain 1, then the second person adds 2 to 0 to obtain 2, then the first person adds 3 to 1 to obtain 4, etc.).

$$3,3 \quad 2,3;3 \quad 2,1;3,3 \quad 1,2;3,3 \quad 1,3;3$$

158. Cyclic equivalence classes (or *necklaces*) of sequences of $n+1$ 1's and n 0's (one sequence from each class is shown below).

$$1111000 \quad 1110100 \quad 1110010 \quad 1101100 \quad 1101010$$

159. Noncrossing partitions of $[n]$, i.e., partitions $\pi = \{B_1,\dots,B_k\} \in \Pi_n$ such that if $a < b < c < d$ and $a,c \in B_i$ and $b,d \in B_j$, then $i = j$.

$$123 \quad 12\text{-}3 \quad 13\text{-}2 \quad 23\text{-}1 \quad 1\text{-}2\text{-}3$$

160. Partitions $\{B_1,\dots,B_k\}$ of $[n]$ such that if the numbers $1,2,\dots,n$ are arranged in order around a circle, then the convex hulls of the blocks $B_1,\dots,B_k$ are pairwise disjoint.

161. Noncrossing Murasaki diagrams (whose definition should be clear from the examples below) with n vertical lines.

162. Noncrossing partitions of some set $[k]$ with $n+1$ blocks, such that any two elements of the same block differ by at least three.

$$1\text{-}2\text{-}3\text{-}4 \quad 14\text{-}2\text{-}3\text{-}5 \quad 15\text{-}2\text{-}3\text{-}4 \quad 25\text{-}1\text{-}3\text{-}4 \quad 16\text{-}25\text{-}3\text{-}4$$

163. Noncrossing partitions of $[2n+1]$ into $n+1$ blocks, such that no block contains two consecutive integers.

$$137\text{-}46\text{-}2\text{-}5 \quad 1357\text{-}2\text{-}4\text{-}6 \quad 157\text{-}24\text{-}3\text{-}6 \quad 17\text{-}246\text{-}3\text{-}5 \quad 17\text{-}26\text{-}35\text{-}4$$

164. *Nonnesting partitions* of $[n]$, i.e., partitions of $[n]$ such that if a, e appear in a block B and b, d appear in a *different* block B' where $a < b < d < e$, then there is a $c \in B$ satisfying $b < c < d$.

$$123 \quad 12\text{-}3 \quad 13\text{-}2 \quad 23\text{-}1 \quad 1\text{-}2\text{-}3$$

(The unique partition of $[4]$ that isn't nonnesting is 14-23.)

165. 231-avoiding partitions of $[n]$, i.e., partitions of $[n]$ such that if they are written with increasing entries in each block and blocks arranged in increasing order of their first entry, then the permutation of $[n]$ obtained by erasing the dividers between the blocks is 231-avoiding.

$$1\text{-}2\text{-}3 \quad 12\text{-}3 \quad 13\text{-}2 \quad 1\text{-}23 \quad 123$$

(The only partition of $[4]$ that isn't 231-avoiding is 134-2.)

166. Equivalence classes of the equivalence relation on the set $S_n = \{(a_1, \ldots, a_n) \in \mathbb{N}^n : \sum a_i = n\}$ generated by $(\alpha, 0, \beta) \sim (\beta, 0, \alpha)$ if β (which may be empty) contains no 0's. For instance, when $n = 7$, one equivalence class is given by

$$\{3010120, 0301012, 1200301, 1012003\}.$$

$$\{300, 030, 003\} \quad \{210, 021\} \quad \{120, 012\} \quad \{201, 102\} \quad \{111\}$$

167. Young diagrams that fit in the shape $(n-1, n-2, \ldots, 1)$.

168. Standard Young tableaux of shape (n, n) (or equivalently, of shape $(n, n-1)$).

123	124	125	134	135
456	356	346	256	246

or

123	124	125	134	135
45	35	34	25	24

169. Pairs (P, Q) of standard Young tableaux of the same shape, each with n squares and at most two rows.

$$(123, 123) \quad \begin{pmatrix} 12 & 12 \\ 3 & 3 \end{pmatrix} \begin{pmatrix} 12 & 13 \\ 3 & 2 \end{pmatrix} \begin{pmatrix} 13 & 12 \\ 2 & 3 \end{pmatrix} \begin{pmatrix} 13 & 13 \\ 2 & 2 \end{pmatrix}$$

170. Standard Young tableaux with at most two rows and with first row of length $n - 1$.

$$\begin{array}{ccccc} 12 & 12 & 13 & 12 & 13 \\ & 3 & 2 & 34 & 24 \end{array}$$

171. Standard Young tableaux with at most two rows and with first row of length n, such that for all i the ith entry of row 2 is not $2i$.

$$\begin{array}{ccccc} 123 & 123 & 124 & 123 & 124 \\ & 4 & 3 & 45 & 35 \end{array}$$

172. Standard Young tableaux of shape $(2n + 1, 2n + 1)$ such that adjacent entries have opposite parity.

$$\begin{bmatrix} 1 & 2 & 3 & 4 & 5 & 6 & 7 \\ 8 & 9 & 10 & 11 & 12 & 13 & 14 \end{bmatrix} \quad \begin{bmatrix} 1 & 2 & 3 & 4 & 5 & 8 & 9 \\ 6 & 7 & 10 & 11 & 12 & 13 & 14 \end{bmatrix}$$

$$\begin{bmatrix} 1 & 2 & 3 & 4 & 5 & 10 & 11 \\ 6 & 7 & 8 & 9 & 12 & 13 & 14 \end{bmatrix} \quad \begin{bmatrix} 1 & 2 & 3 & 6 & 7 & 8 & 9 \\ 4 & 5 & 10 & 11 & 12 & 13 & 14 \end{bmatrix}$$

$$\begin{bmatrix} 1 & 2 & 3 & 6 & 7 & 10 & 11 \\ 4 & 5 & 8 & 9 & 12 & 13 & 14 \end{bmatrix}$$

173. Column-strict plane partitions of shape $(n-1, n-2, \ldots, 1)$, such that each entry in the ith row is equal to $n - i$ or $n - i + 1$.

$$\begin{array}{ccccc} 33 & 33 & 32 & 32 & 22 \\ 2 & 1 & 2 & 1 & 1 \end{array}$$

174. Arrays

$$\begin{pmatrix} a_1 & a_2 & \cdots & a_{r-1} & a_r \\ b_1 & b_2 & \cdots & b_{r-1} & \end{pmatrix}$$

of integers, for some $r \geq 1$, such that $a_i > 0$, $b_i \geq 0$, $\sum a_i = n$, and $b_i < a_i + b_{i-1}$ for $1 \leq i \leq r - 1$ (setting $b_0 = 0$).

$$\begin{pmatrix} 1 & 1 & 1 \\ 0 & 0 & \end{pmatrix} \begin{pmatrix} 2 & 1 \\ 0 & \end{pmatrix} \begin{pmatrix} 2 & 1 \\ 1 & \end{pmatrix} \begin{pmatrix} 1 & 2 \\ 0 & \end{pmatrix} \begin{pmatrix} 3 \end{pmatrix}$$

175. Plane partitions with largest part at most two and contained in a rectangle of perimeter $2(n-1)$ (including degenerate $0 \times (n-1)$ and $(n-1) \times 0$ rectangles)

176. Convex subsets S of the poset $\mathbb{Z} \times \mathbb{Z}$, up to translation by a diagonal vector (m,m), such that if $(i,j) \in S$ then $0 < i - j < n$.

$$\emptyset \quad \{(1,0)\} \quad \{(2,0)\} \quad \{(1,0),(2,0)\} \quad \{(2,0),(2,1)\}$$

177. Linear extensions of the poset $\mathbf{2} \times \mathbf{n}$, where in general $\mathbf{m}$ denotes an m-element chain and $\times$ denotes direct (or Cartesian) product.

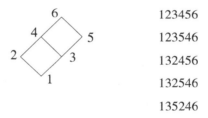

123456

123546

132456

132546

135246

178. Order ideals of $\text{Int}(\mathbf{n-1})$, the poset of intervals of the $(n-1)$-element chain $\mathbf{n-1}$.

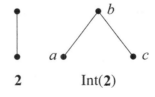

$$\mathbf{2} \qquad \text{Int}(\mathbf{2})$$

$$\emptyset, \ a, \ c, \ ac, \ abc$$

179. Order ideals of the poset A_n obtained from the poset $(\mathbf{n-1}) \times (\mathbf{n-1})$ by adding the relations $(i,j) < (j,i)$ if $i > j$ (see Figure 2.2 for the Hasse diagram of A_4).

$$\emptyset \quad \{11\} \quad \{11,21\} \quad \{11,21,12\} \quad \{11,21,12,22\}$$

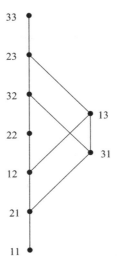

Figure 2.2. A poset with $C_4 = 14$ order ideals.

180. Nonisomorphic n-element posets with no induced subposet isomorphic to $\mathbf{2+2}$ or $\mathbf{3+1}$, where $+$ denotes disjoint union.

181. Nonisomorphic $(n+1)$-element posets that are a union of two chains and that are not a (nontrivial) ordinal sum, rooted at a minimal element.

182. Nonisomorphic n-element posets with no induced subposet isomorphic to $\mathbf{2+2}$ or the fence Z_4 (the poset defined by the relations $a < b > c < d$).

183. Nonisomorphic $2(n+1)$-element posets that are a union of two chains, that are not a (nontrivial) ordinal sum, and that have a nontrivial automorphism.

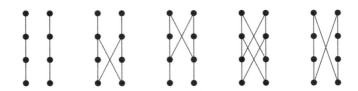

184. Natural partial orderings $<_P$ of $[n]$ such that if $i <_P k$ and $i <_{\mathbb{Z}} j <_{\mathbb{Z}} k$, then $i <_P j$.

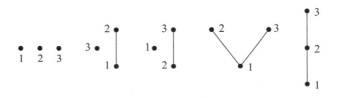

185. Natural partial orderings $<_P$ of $[n]$ such that if $i <_{\mathbb{Z}} j <_P k$ then $i <_P k$, and dually if $i >_{\mathbb{Z}} j >_P k$ then $i >_P k$.

186. Sets S of n non-identity permutations in $\mathfrak{S}_{n+1}$ such that every pair (i,j) with $1 \le i < j \le n$ is an inversion of exactly one permutation in S.

$$\{1243, 2134, 3412\} \quad \{1324, 2314, 4123\} \quad \{2134, 3124, 4123\}$$

$$\{1324, 1423, 2341\} \quad \{1243, 1342, 2341\}$$

187. Relations R on $[n]$ that are reflexive (iRi), symmetric ($iRj \Rightarrow jRi$), and such that if $1 \le i < j < k \le n$ and iRk, then iRj and jRk (in the example below we write ij for the pair (i,j), and we omit the pairs ii).

$$\emptyset \quad \{12, 21\} \quad \{23, 32\} \quad \{12, 21, 23, 32\} \quad \{12, 21, 13, 31, 23, 32\}$$

188. Permutations $w \in W(\mathfrak{S}_n)$, the weak (Bruhat) order of $\mathfrak{S}_n$ (as defined in [64, Exercise 3.185]), for which the interval $[\hat{0}, w]$ is a distributive lattice

$$123 \qquad 132 \qquad 213 \qquad 231 \qquad 312$$

189. Subsets S of the order ideal $I_n = \{(i,j) : i+j \le n-2\}$ of $\mathbb{N} \times \mathbb{N}$ such that for all $(i,j) \in I$ we have $\#(S \cap V_{(i,j)}) \le n-1-i-j$, with equality if and only if $(i,j) \in S$ (where $V_{(i,j)} = \{(h,k) \in I_n : (h,k) \ge (i,j)\}$).

$$\emptyset \quad \{10\} \quad \{01\} \quad \{00,10\} \quad \{00,01\}$$

190. Isomorphism classes of ordered pairs (S,R) of binary relations on $[n]$ such that S and R are irreflexive (i.e., don't contain any (i,i)) and transitive, and such that $\overline{R} \cup \overline{S} = [n]^2 - \{(i,i) : i \in [n]\}$, $\overline{R} \cap \overline{S} = \emptyset$, and $S \circ R \subseteq R$. Here

$$\overline{T} = T \cup \{(j,i) : (i,j) \in T\}$$
$$S \circ R = \{(i,k) : \exists j \ (i,j) \in S, \ (j,k) \in R\}.$$

Two pairs (S,R) and (S',R') are considered isomorphic if there is a bijection $f : [n] \to [n]$ inducing bijections $S \to S'$ and $R \to R'$.

$$(\emptyset, \{12, 13, 23\}) \quad (\{12\}, \{13, 23\}) \quad (\{23\}, \{12, 13\})$$
$$(\{13, 23\}, \{12\}) \quad (\{12, 13, 23\}, \emptyset)$$

191. Ways to stack coins in the plane, the bottom row consisting of n consecutive coins.

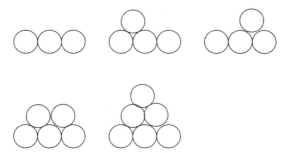

192. n-element multisets on $\mathbb{Z}/(n+1)\mathbb{Z}$ whose elements sum to 0.

$$000 \quad 013 \quad 022 \quad 112 \quad 233$$

193. $(n+1)$-element multisets on $\mathbb{Z}/n\mathbb{Z}$ whose elements sum to 0.

$$0000 \qquad 0012 \qquad 0111 \qquad 0222 \qquad 1122$$

194. n-element subsets S of $\mathbb{N} \times \mathbb{N}$ such that if $(i,j) \in S$ then $i \geq j$ and there is a lattice path from $(0,0)$ to (i,j) with steps $(0,1)$, $(1,0)$, and $(1,1)$ that lies entirely inside S.

$$\{(0,0),(1,0),(2,0)\} \quad \{(0,0),(1,0),(1,1)\} \quad \{(0,0),(1,0),(2,1)\}$$

$$\{(0,0),(1,1),(2,1)\} \quad \{(0,0),(1,1),(2,2)\}$$

195. Regions into which the cone $x_1 \geq x_2 \geq \cdots \geq x_n$ in $\mathbb{R}^n$ is divided by the hyperplanes $x_i - x_j = 1$, for $1 \leq i < j \leq n$ (the diagram below shows the situation for $n = 3$, intersected with the hyperplane $x_1 + x_2 + x_3 = 0$).

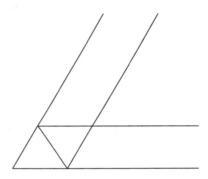

196. Bounded regions into which the cone $x_1 \geq x_2 \geq \cdots \geq x_{n+1}$ in $\mathbb{R}^{n+1}$ is divided by the hyperplanes $x_i - x_j = 1$, $1 \leq i < j \leq n+1$ (compare item 195, which illustrates the case $n = 2$ of the present item).

197. Positive integer sequences $a_1, a_2, \ldots, a_{n+2}$ for which there exists an integer array (necessarily with $n + 1$ rows),

1	1	1	$\cdots$	1	1	1	$\cdots$	1	1
a_1	a_2	a_3	$\cdots$	a_{n+2}	a_1	a_2	$\cdots$	a_{n-1}	
	b_1	b_2	b_3	$\cdots$	b_{n+2}	b_1	$\cdots$	b_{n-2}	

$$\vdots$$

	r_1	r_2	r_3	$\cdots$	r_{n+2}	r_1			
		1	1	1	$\cdots$	1			

(2.1)

such that any four neighboring entries in the configuration
$$\begin{matrix} & r & \\ s & & t \\ & u & \end{matrix}$$
satisfy $st = ru + 1$. An example of such an array for $(a_1, \ldots, a_8) = (1,3,2,1,5,1,2,3)$, necessarily unique, is given by Figure 2.3.

$$12213 \quad\quad 22131 \quad\quad 21312 \quad\quad 13122 \quad\quad 31221$$

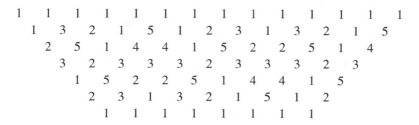

Figure 2.3. The frieze pattern corresponding to the sequence $(1,3,2,1,5,1,2,3)$.

198. Ways to choose a lattice path of length $n-1$ from some point $(0,j)$ to some point $(i,0)$, where $i,j \geq 0$, with steps $(1,0)$ and $(0,-1)$, and then coloring either red or blue some of the unit coordinate squares (cells) below the path and in the first quadrant, satisfying:
   • There is no colored cell below a red cell.
   • There is no colored cell to the left of a blue cell.
   • Every uncolored cell is either below a red cell or to the left of a blue cell.

199. n-element subsets S of $\mathbb{N} \times \mathbb{N}$ satisfying the following three conditions: (a) there are no gaps in row 0, i.e., if $(k,0) \in S$ and $k > 0$, then $(k-1,0) \in S$; (b) every nonempty column has a row 0 element, i.e., if $(k,i) \in S$ then $(k,0) \in S$; and (c) the number of gaps in column k is at most the number of elements in column $k+1$, i.e., if $G_k = \{(k,i) \notin S : (k,j) \in S \text{ for some } j > i\}$ and $S_k = \{i : (k,i) \in S\}$, then $\#S_k \geq \#G_k$.

$\{00,01,02\} \ \{00,02,10\} \ \{00,10,11\} \ \{00,01,10\} \ \{00,10,20\}$

200. Triples (A,B,C) of pairwise disjoint subsets of $[n-1]$ such that $\#A = \#B$ and every element of A is less than every element of B.

$(\emptyset,\emptyset,\emptyset) \quad (\emptyset,\emptyset,1) \quad (\emptyset,\emptyset,2) \quad (\emptyset,\emptyset,12) \quad (1,2,\emptyset)$

201. Subsets S of $\mathbb{N}$ such that $0 \in S$ and such that if $i \in S$ then $i+n, i+n+1 \in S$.

$\mathbb{N}, \quad \mathbb{N}-\{1\}, \quad \mathbb{N}-\{2\}, \quad \mathbb{N}-\{1,2\}, \quad \mathbb{N}-\{1,2,5\}$

202. Maximal chains $\emptyset = S_0 \subset S_1 \subset \cdots \subset S_n = [n]$ of subsets of $[n]$ such that $S_i - S_{i-1} = \{m\}$ if and only if m belongs to the rightmost maximal set of consecutive integers contained in S_i.

$$\emptyset \subset 1 \subset 12 \subset 123, \quad \emptyset \subset 2 \subset 12 \subset 123, \quad \emptyset \subset 1 \subset 13 \subset 123$$

$$\emptyset \subset 2 \subset 23 \subset 123, \quad \emptyset \subset 3 \subset 23 \subset 123$$

203. Ways to write $(1,1,\ldots,1,-n) \in \mathbb{Z}^{n+1}$ as a sum of vectors $e_i - e_{i+1}$ and $e_j - e_{n+1}$, without regard to order, where e_k is the kth unit coordinate vector in $\mathbb{Z}^{n+1}$.

$$(1,-1,0,0) + 2(0,1,-1,0) + 3(0,0,1,-1)$$
$$(1,0,0,-1) + (0,1,-1,0) + 2(0,0,1,-1)$$
$$(1,-1,0,0) + (0,1,-1,0) + (0,1,0,-1) + 2(0,0,1,-1)$$
$$(1,-1,0,0) + 2(0,1,0,-1) + (0,0,1,-1)$$
$$(1,0,0,-1) + (0,1,0,-1) + (0,0,1,-1)$$

204. Horizontally convex polyominoes (as defined in [64, §4.7.5]) of width (number of columns) $n+1$ such that each row begins strictly to the right of the beginning of the previous row and ends strictly to the right of the end of the previous row.

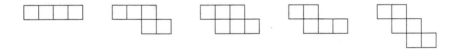

205. Tilings of the staircase shape $(n, n-1, \ldots, 1)$ with n rectangles.

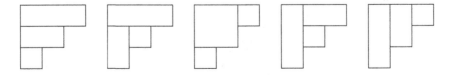

206. Complete matchings of the triangular benzenoid graph T_{n-1} of order $n-1$. The graph T_n is a planar graph whose bounded regions are hexagons, with i hexagons in row i (from the top) and n rows in all, as illustrated for $n = 4$ in Figure 2.4.

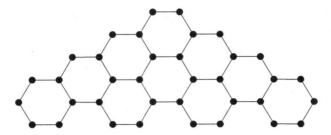

Figure 2.4. The triangular benzenoid graph T_4.

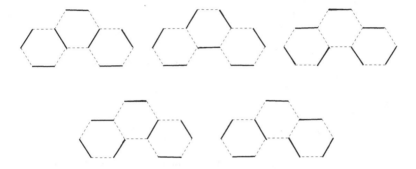

207. n-tuples $(a_1, \ldots a_n)$ of positive integers such that the tridiagonal matrix

$$
\begin{bmatrix}
a_1 & 1 & 0 & 0 & \cdot & \cdot & \cdot & 0 & 0 \\
1 & a_2 & 1 & 0 & \cdot & \cdot & \cdot & 0 & 0 \\
0 & 1 & a_3 & 1 & \cdot & \cdot & \cdot & 0 & 0 \\
 & & & & \cdot & & & & \\
 & & & & \cdot & & & & \\
 & & & & \cdot & & & & \\
0 & 0 & 0 & 0 & \cdot & \cdot & \cdot & a_{n-1} & 1 \\
0 & 0 & 0 & 0 & \cdot & \cdot & \cdot & 1 & a_n
\end{bmatrix}
$$

is positive definite with determinant one.

131   122   221   213   312

208. $n \times n$ $\mathbb{N}$-matrices $M = (m_{ij})$ where $m_{ij} = 0$ unless $i = n$ or $i = j$ or $i = j - 1$, with row and column sum vector $(1, 2, \ldots, n)$.

$$
\begin{bmatrix} 1 & 0 & 0 \\ 0 & 2 & 0 \\ 0 & 0 & 3 \end{bmatrix}
\quad
\begin{bmatrix} 0 & 1 & 0 \\ 0 & 1 & 1 \\ 1 & 0 & 2 \end{bmatrix}
\quad
\begin{bmatrix} 1 & 0 & 0 \\ 0 & 1 & 1 \\ 0 & 1 & 2 \end{bmatrix}
\quad
\begin{bmatrix} 1 & 0 & 0 \\ 0 & 0 & 2 \\ 0 & 2 & 1 \end{bmatrix}
\quad
\begin{bmatrix} 0 & 1 & 0 \\ 0 & 0 & 2 \\ 1 & 1 & 1 \end{bmatrix}
$$

209. $(n+1) \times (n+1)$ alternating sign matrices (i.e., matrices with entries $0, \pm 1$ such that every row and column sum is 1, and the nonzero entries of each row alternate between 1 and -1) such that the rightmost 1 in row $i+1 \geq 2$ occurs to the right of the leftmost 1 in row i.

$$\begin{bmatrix} 1 & 0 & 0 & 0 \\ 0 & 1 & 0 & 0 \\ 0 & 0 & 1 & 0 \\ 0 & 0 & 0 & 1 \end{bmatrix} \begin{bmatrix} 0 & 1 & 0 & 0 \\ 1 & -1 & 1 & 0 \\ 0 & 1 & 0 & 0 \\ 0 & 0 & 0 & 1 \end{bmatrix} \begin{bmatrix} 0 & 1 & 0 & 0 \\ 0 & 0 & 1 & 0 \\ 1 & -1 & 0 & 1 \\ 0 & 1 & 0 & 0 \end{bmatrix}$$

$$\begin{bmatrix} 0 & 0 & 1 & 0 \\ 1 & 0 & -1 & 1 \\ 0 & 1 & 0 & 0 \\ 0 & 0 & 1 & 0 \end{bmatrix} \begin{bmatrix} 0 & 0 & 1 & 0 \\ 0 & 1 & -1 & 1 \\ 1 & -1 & 1 & 0 \\ 0 & 1 & 0 & 0 \end{bmatrix}$$

210. Number of distinct terms (monomials) appearing in the expansion of $\prod_{i=1}^{n}(x_1 + x_2 + \cdots + x_i)$.

$$x(x+y)(x+y+z) = x^3 + 2x^2 y + xy^2 + x^2 z + xyz$$

211. Number of words of length $2n$ (where the empty word 1 has length 0) in the noncommutative power series F in the variables x, y satisfying $F = 1 + xFyF = 1 + xy + xxyy + xyxy + \cdots$.

$$xxxyyy \quad xxyxyy \quad xxyyxy \quad xyxxyy \quad xyxyxy$$

212. Number of distinct volumes of the sets

$$X_w = \{x \in B : x_{w(1)} < x_{w(2)} < \cdots < x_{w(n)}\},$$

where we fix some sequence $0 < a_1 < a_2 < \cdots < a_n$ and define

$$B = [0, a_1] \times [0, a_2] \times \cdots \times [0, a_n] \subset \mathbb{R}^n.$$

$$\mathrm{vol}(X_{231}) = \mathrm{vol}(X_{321}), \ \mathrm{vol}(X_{312}), \ \mathrm{vol}(X_{213}), \ \mathrm{vol}(X_{132}), \ \mathrm{vol}(X_{123})$$

213. Facets of the cone of all functions $F \colon 2^{[n]} \to \mathbb{R}$ such that $\sum_{S \subseteq [n]} \alpha_P(S) F(S) \geq 0$ for all graded posets P of rank $n+1$ with $\hat{0}$ and $\hat{1}$, where α_P denotes the flag f-vector of P [64, §3.13].

$$F(\emptyset) \geq 0, \ F(1) - F(\emptyset) \geq 0, \ F(2) - F(\emptyset) \geq 0, \ F(1,2) - F(1) \geq 0,$$
$$F(1,2) - F(2) \geq 0$$

214. Number of elements, when written with the maximum possible number of x's, with n x's in the free near semiring N_1 on the generator x. (A *near semiring* is a set with two associative operations $+$ and $\cdot$ such that $(a+b)\cdot c = a\cdot c + b\cdot c$. Neither operation is required to be commutative. Thus $(x+x)\cdot x$ contributes to C_4 since it can be written as $x\cdot x + x\cdot x$.)

$$x+x+x \qquad x+x\cdot x \qquad x\cdot x+x \qquad x\cdot(x+x) \qquad x\cdot x\cdot x$$

3

Bijective Solutions

It would require vastly more space to discuss thoroughly all the known interconnections among these problems. We will content ourselves with some brief comments and references that should serve as a means of further exposure to "Catalan disease" or "Catalania" (= Catalan mania).

First note that items 1, 3, 4, 6, 24, 25, and 77 are covered by Theorem 1.5.1.

2. The total number of triangulations using the triangle with vertices $1, i, i + 1$ is easily seen to equal $C_{i-2}C_{n-i+1}$. The proof follows from the fundamental recurrence (1.1) and checking the initial condition $n = 0$ or $n = 1$. This item is due to L. Shapiro, private communication, April 9, 2014. Shapiro calls a triangle being counted a *wedge*.

5. Do a depth-first search, recording 1 when a left edge is first encountered, and recording -1 when a right edge is first encountered. This gives a bijection with item 77. Note also that when all endpoints are removed (together with the incident edges), we obtain the trees of item 4.

7. When the root is removed, we obtain the trees of item 5. See also Klarner [32].

8. The bijection between parts (iii) and (vi) of Theorem 1.5.1 gives a bijection between the present problem and item 26. An elegant bijection with item 6 was given by F. Bernhardt, private communication, 1996.

9. Traverse the tree in preorder. When going down an edge (i.e., away from the root), record 1 if this edge goes to the left or straight down, and record -1 if this edge goes to the right. This gives a bijection with item 98. This item is due to E. Deutsch.

10. Write

$$E(x) = \sum_{n \geq 0} C_{2n}x^{2n} = \frac{1}{2}(C(x) + C(-x))$$

$$O(x) = \sum_{n \geq 0} C_{2n+1}x^{2n+1} = \frac{1}{2}(C(x) - C(-x)).$$

The proof follows from the generating function identity

$$C(x)^2 = \frac{C(x) - 1}{x}$$

$$= \sum_{k \geq 0} x^k \sum_{i=0}^{k} O(x)^i 2^{k-i} E(x)^{k-i}$$

$$= \frac{1}{(1 - 2xE(x))(1 - xO(x))}.$$

This result is due to L. Shapiro, private communication, December 26, 2001, who raises the question of giving a simple bijective proof. In a preprint entitled "Catalan trigonometry" he gives a simple bijective proof of the related identity

$$O(x) = x(O(x)^2 + E(x)^2)$$

and remarks that there is a similar proof of

$$E(x) = 1 + 2xE(x)O(x).$$

For a further identity of this nature, see Problem A33.

11. Given a tree T of the type being counted, start with an up step U. Continue with the usual preorder bijection between plane trees and Dyck paths (proof of Theorem 1.5.1), except that at a blue leaf take the steps DU rather than UD. Finish with a down step D. This gives a bijection with the Dyck paths of item 25. This result is due to L. Shapiro, private communication, May 24, 2002.

12. Traverse the tree in preorder. Replace an edge that is traversed for the first time (i.e., away from the root) by $1, -1$ if its vertex v furthest from the root has no siblings, by 1 if v is a left child, and by -1 if v is a right child, yielding a bijection with the ballot sequences of item 77. This item is due to E. Deutsch, private communication, February 27, 2007.

13. Let $B(x) = xC(x)$, so $B(x) = x + B(x)^2$. Hence $B(x) = x + xB(x) + B(x)^3$, from which the proof follows. We can also give a recursively

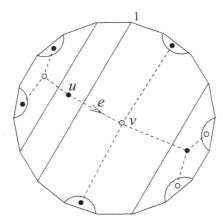

Figure 3.1. A bicolored edge-rooted plane tree associated with nonintersecting chords.

defined bijection between the binary trees of item 4 and the trees being counted, basically by decomposing a binary tree into three subtrees: the left subtree of the root, the left subtree of the right subtree of the root, and the right subtree of the right subtree of the root. This item is due to I. Gessel, private communication, May 15, 2008.

14. Take n nonintersecting chords joining $2n$ points $1, 2, \ldots, 2n$ on the circumference of a circle as in item 59. Let T be the "interior dual graph" of these chords together with additional edges between every two consecutive vertices, as illustrated in Figure 3.1. Thus T is an unrooted plane tree with $n + 1$ vertices. Let the root edge e of T be the edge crossing the chord containing vertex 1. Let u, v be the vertices of e, and color u black (respectively, white) if in walking from u to v vertex 1 is on the left (respectively, right). This procedure sets up a bijection between item 59 and the present item.

    The result of this item is due to M. Bóna, M. Bousquet, G. Labelle, and P. Leroux, *Adv. in Appl. Math.* **24** (2000), 22–56 (the case $m = 2$ of Theorem 10). A bijection with the noncrossing partitions of item 159 similar to the bijection above was later given by D. Callan and L. Smiley, math.CO/0510447.

15. We obtain a bijection with plane trees (item 6) as follows. First, erase the leaf labels and transfer the remaining labels from a vertex to the parent edge. Then process the labeled edges from bottom to top: for an edge with label i, transfer the $i - 1$ rightmost subtrees of the child vertex, preserving order, to become the rightmost subtrees

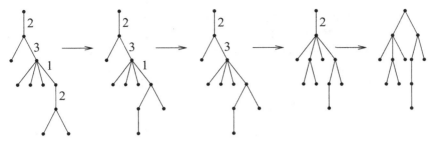

Figure 3.2. An example of the bijection of item 15.

of the parent vertex, and then erase the label. See Figure 3.2 for an example.

Trees satisfying conditions (i) and (ii) are known as $\beta(1,0)$-*trees*, due to Claesson, Kitaev, and Steingrímsson, *Adv. in Appl. Math.* **42** (2009), 313–328. The additional condition (iii) and bijection above are due to D. Callan, private communication, February 19, 2008.

16. Associate with a tree T being counted a sequence $(a_1,\ldots,a_n)$, where $a_i = \operatorname{outdeg}(i) - 1$. This sets up a bijection with item 82 (D. Callan, private communication, May 28, 2010).

17. Let $F(x,t) = \sum_T t^{s(T)} x^{v(T)}$, where the sum is over all (nonempty) plane trees, $v(T)$ is the number of vertices of T, and $s(T)$ is the number of vertices with exactly one child on the leftmost path from the root. Thus $F(x,1) = xC(x) = \frac{1}{2}(1 - \sqrt{1-4x})$, and we want the coefficients of $\frac{\partial}{\partial t} F(x,t)|_{t=1}$. A simple combinatorial argument gives

$$F(x,t) = x + xty + \frac{xF(x,1)y}{1 - F(x,1)}.$$

It is now routine to compute

$$\frac{\partial}{\partial t} F(x,t)|_{t=1} = xC(x) - x,$$

and the proof follows. This result is due to L. Shapiro, private communication dated March 17, 2014. Shapiro also has a bijective proof.

18. In the generating function $F(x,t)$ of the previous item, put $t = 2$ to get $C(x) - 1$, and the proof follows. This result is due to L. Shapiro, private communication dated March 17, 2014. Shapiro also has a bijective proof.

19. A bijection with item 67 was given by J. S. Kim, *SIAM J. Discrete Math.* **25**(1) (2011), 447–461. See also S. Forcey, M. Kafashan, M. Maleki, and M. Strayer, *J. Integer Seq.* **16** (2013), article 13.5.3.

20. See the numbers U_m of S. J. Cyvin, J. Brunvoll, E. Brendsdal, B. N. Cyvin, and E. K. Lloyd, *J. Chem. Inf. Comput. Sci.* **35** (1995), 743–751.

21. The labeled trees being counted are just the plane trees of item 6 with the preorder (or depth-first) labeling. A more complicated bijection with Dyck paths appears in Section 2.3 of the paper A. Asinowski and T. Mansour, *Europ. J. Combinatorics* **29** (2008), 1262–1279.

22. List the non-endpoint vertices $v_1, \ldots, v_r$ in increasing order along with their degrees (number of children) $d_1, \ldots, d_r$. Then the pairs $((d_1, d_2, \ldots, d_r), (v_2 - v_1, v_3 - v_2, \ldots, v_r - v_{r-1}, n + 1 - v_r))$ are in bijection with item 109. For instance, the five trees listed in the statement of this item correspond to the pairs $(3, 3)$, $(21, 12)$, $(21, 21)$, $(12, 12)$, $(111, 111)$ in that order. This item is due to D. Callan, private communication, September 1, 2007. Note the noncrossing/nonnesting analogy between the pairs of items $(62, 63)$, $(159, 164)$, $(61, 64)$, and $(21, 22)$.

23. *First solution.* Fix a permutation $u \in \mathfrak{S}_3$, and let $T(u)$ be the set of all u-avoiding permutations (as defined in Exercise 6.39(l)) in all $\mathfrak{S}_n$ for $n \geq 1$. Partially order $T(u)$ by setting $v \leq w$ if v is a subsequence of w (when v and w are written as words). One checks that $T(u)$ is isomorphic to the tree T. Moreover, the vertices in $T(u)$ of height n consist of the u-avoiding permutations in $\mathfrak{S}_{n+1}$. The proof then follows from items 115 and 116.

Second solution. Label each vertex by its degree. A saturated chain from the root to a vertex at level $n - 1$ is thus labeled by a sequence $(b_1, b_2, \ldots, b_n)$. Set $a_i = i + 2 - b_i$. This sets up a bijection between level $n - 1$ and the sequences $(a_1, \ldots, a_n)$ of item 78.

Third solution. Let $f_n(x) = \sum_v x^{\deg(v)}$, summed over all vertices of T of height $n - 1$. Thus $f_1(x) = x^2, f_2(x) = x^2 + x^3, f_3(x) = x^4 + 2x^3 + 2x^2$, etc. Set $f_0(x) = x$. The definition of T implies that we get $f_{n+1}(x)$ from $f_n(x)$ by substituting $x^2 + x^3 + \cdots + x^{k+1} = x^2(1 - x^k)/(1 - x)$ for x^k. Thus

$$f_{n+1}(x) = \frac{x^2(f_n(1) - f_n(x))}{1 - x}, \quad n \geq 0.$$

Setting $F(x,t) = \sum_{n \geq 0} F_n(x)t^n$, there follows

$$\frac{x - F(x,t)}{t} = \frac{x^2}{1-x}(F(1,t) - F(x,t)).$$

Hence

$$F(x,t) = \frac{x - x^2 + x^2 t F(1,t)}{1 - x + x^2 t}.$$

Now use Problem A34.

The trees $T(u)$ were first defined by J. West, *Discrete Math.* **146** (1995), 247–262 (see also *Discrete Math.* **157** (1996), 363–374) as in the first solution above, and are called *generating trees*. West then presents the labeling argument of the second solution, thereby giving new proofs of items 115 and 116. For further information on generating trees, see C. Banderier, M. Bousquet-Mélou, A. Denise, P. Flajolet, D. Gardy, and D. Gouyou-Beauchamps, *Discrete Math.* **246** (2002), 29–55.

26. Let $A(x) = x + x^3 + 2x^4 + 6x^5 + \cdots$ (respectively, $B(x) = x^2 + x^3 + 3x^4 + 8x^5 + \cdots$) be the generating function for Dyck paths from $(0,0)$ to $(2n,0)$ $(n > 0)$ such that the path only touches the x-axis at the beginning and end, and the number of steps $(1,-1)$ at the end of the path is odd (respectively, even). Let $C(x) = 1 + x + 2x^2 + 5x^3 + \cdots$ be the generating function for all Dyck paths from $(0,0)$ to $(2n,0)$, so the coefficients are Catalan numbers by item 25. It is easy to see that $A = x(1 + CB)$ and $B = xCA$. (Also $C = 1/(1 - A - B)$, though we don't need this fact here.) Solving for A gives $A = x/(1 - x^2 C^2)$. The generating function we want is $1/(1 - A)$, which simplifies (using $1 + xC^2 = C$) to $1 + xC$, and the proof follows. This result is due to E. Deutsch, private communication, 1996. Presumably there is a reasonably simple bijective proof.

27. This result is due to P. Peart and W. Woan, *J. Integer Seq.* **4** (2001), article 01.1.3. The authors give a generating function proof and a simple bijection with item 25.

28. Add one further up step and then down steps until reaching $(2n,0)$. This gives a bijection with the Dyck paths of item 25. This result is due to E. Deutsch.

29. Deleting the first UD gives a bijection with Dyck paths of length $2n$ (item 25). This result is due to D. Callan, private communication, November 3, 2004.

30. The generating function for Dyck paths with no peaks at height 1 (or *hill-free* Dyck paths) is

$$F(x) = \frac{1}{1 - x^2 C(x)^2}.$$

The generating function for hill-free Dyck paths with leftmost peak at height 2 is $x^2 C(x) F(x)$. The generating function for hill-free Dyck paths with leftmost peak at height 3 is $x^3 C(x)^2 F(x)$. But it can be checked that

$$x^2 C(x) F(x) + x^3 C(x)^2 F(x) = x(C(x) - 1),$$

and the proof follows. This result is due to E. Deutsch, private communication, December 30, 2006. Is there a nice bijective proof?

31. See D. Callan, math.CO/0511010.

32. Given such a path, delete the first and last steps and every valley (down step immediately followed by an up step). This is a bijection to Dyck paths of length $2n$ (item 25) because the paths being counted necessarily have $n - 1$ valleys. Another nice bijection is the following. Traverse the plane binary trees of item 6 in preorder. To each node of degree r, associate r up steps followed by one down step, except that the last leaf is ignored. We then get a Dyck path such that every *ascent* has length 2, clearly in bijection with those for which every *descent* has length 2. This item is due independently to D. Callan (first proof above, private communication, September 1, 2007) and E. Deutsch (second proof, private communication, September 6, 2007).

33. In the two-colored Motzkin paths of item 40 replace the step $(1, 1)$ with the sequence of steps $(1, 1) + (1, 1) + (1, -1)$, the step $(1, -1)$ with $(1, 1) + (1, -1) + (1, -1)$, the red step $(1, 0)$ with $(1, 1) + (1, -1)$, and the blue step $(1, 0)$ with $(1, 1) + (1, 1) + (1, -1) + (1, -1)$.

34. Traverse the trees of item 12 in preorder. Replace a vertex of degree d encountered for the first time with d 1's followed by -1, except do nothing for the last leaf. This gives a bijection with the present item. This item is due to E. Deutsch, private communication, February 27, 2007.

35. For each Dyck path of length $2n$ with at least one valley, insert a UD after the first DU. For the unique Dyck path of length $2n$ without a valley (viz., $U^n D^n$), insert U at the beginning and D at the end. This

sets up a bijection with item 25. This result is due to E. Deutsch, private communication, November 19, 2005.

36. Every Dyck path P with at least two steps has a unique factorization $P = XYZ$ such that Y is a Dyck path (possibly with 0 steps), length$(X) =$ length(Z), and XZ is a Dyck path (with at least two steps) of the type being counted. Hence if $f(n)$ is the number of Dyck paths being counted and $F(x) = \sum_{n \geq 1} f(n-1)x^n$, then

$$C(x) = 1 + F(x)C(x).$$

It follows that $F(x) = xC(x)$, so $f(n) = C_n$ as desired. This result is due to S. Elizalde, private communication, September 2002.

37. To obtain a bijection with the Dyck paths of item 25, add a $(1,1)$ step immediately following a path point $(m,0)$ and a $(1,-1)$ step at the end of the path (R. Sulanke, private communication from E. Deutsch, February 4, 2002).

38. To obtain a bijection with item 37, contract a region under a peak of height one to a point (E. Deutsch and R. Sulanke, private communication from E. Deutsch, February 4, 2002). Deutsch points out (private communication, October 15, 2008) that this item and the previous are special cases of the following result. Fix a Dyck path D of length $2p \leq 2n$. Then the number of occurrences of D beginning at height 0 among all Dyck paths of length $2n$ is C_{n+1-p}. There is a straightforward proof using generating functions.

39. These decompositions are equivalent to the *centered tunnels* of item 36 and are due to S. Elizalde, *Electron. J. Combin.* **18** (2011), P29. Elizalde and E. Deutsch, *Ann. Combinatorics* **7** (2003), 281–297, give a (length-preserving) bijection from Dyck paths to themselves that take centered tunnels into hills. Now use item 38.

40. Replace a step $(1,1)$ with $1,1$, a step $(1,-1)$ with $-1,-1$, a red step $(1,0)$ with $1,-1$, a blue step $(1,0)$ with $-1,1$, and adjoin an extra 1 at the beginning and -1 at the end. This gives a bijection with item 77 (suggested by R. Chapman). The paths being enumerated are called *two-colored Motzkin paths*. See, for instance, E. Barcucci, A. Del Lungo, E. Pergola, and R. Pinzani, *Lecture Notes in Comput. Sci.* **959**, Springer, Berlin, 1995, pp. 254–263.

41. Replace a step $(1,1)$ with $1,1$, a step $(1,-1)$ with $-1,-1$, a red step $(1,0)$ with $-1,1$, and a blue step $(1,0)$ with $1,-1$, to get a bijection with item 77. This item is due to E. Deutsch, private communication, December 28, 2006.

42. Replace each level step in such a path with $1, -1$, each up step with 1, and each down step with -1 to obtain a bijection with the ballot sequences of item 77. This item is due to E. Deutsch, private communication, February 27, 2007.

43. Let π be a noncrossing partition of $[n]$. Denote the steps in a Motzkin path by U (up), D (down) and L (level). In the sequence $1, 2, \ldots, n$, replace the smallest element of a nonsingleton block of π with the two steps LU. Replace the largest element of a nonsingleton block of π with DL. Replace the element of a singleton block with LL. Replace an element that is neither the smallest nor largest element of its block with DU. Remove the first and last terms (which are always L). For instance, if $\pi = 145$–26–3, then we obtain $ULULLDUDLD$. This sets up a bijection with noncrossing partitions (item 159). E. Deutsch (private communication, September 16, 2004) has also given a simple bijection with the Dyck paths of item 25. The bijection given here is a special case of a bijection appearing in W. Chen, E. Deng, R. Du, R. Stanley, and C. Yan, *Trans. Amer. Math. Soc.* **359** (2007), 1555–1575.

44. This result is equivalent to the identity

$$C_n = \sum_{k=0}^{n-1} \binom{n-1}{k} M_k,$$

which can be proved in various ways. See the paragraph preceding Corollary 5.5 of A. Claesson and S. Linusson, *Proc. Amer. Math. Soc.* **139** (2011), 435–449.

45. Replace each level step with an up step followed by a down step to obtain a bijection with the Dyck paths of item 25. This item and the two following are due to L. Shapiro, private communication, August 19, 2005.

46. Add an up step at the beginning and a down step at the end. Then replace each level step at height one with a down step followed by an up step, and replace all other level steps with an up step followed by a down step. We obtain a bijection with Dyck paths.

47. Exactly the same rule as in the solution to item 45 gives a bijection with Dyck paths.

48. Replace each level step with a down step followed by an up step, prepend an up step, and append a down step to obtain a bijection with Dyck paths of item 25 (E. Deutsch, private communication, December 20, 2006).

49. Let U, D, L denote an up step, a down step, and a level (horizontal) step, respectively. Every nonempty Schröder path has uniquely the form LA or $UBDC$, where A, B, C are Schröder paths. Define recursively $\emptyset' = \emptyset$, $(LA)' = LA'$, $(UBDC)' = UC'DB'$. This defines an involution on Schröder paths from $(0,0)$ to $(2n-2,0)$ that interchanges valleys DU and double rises UU. Hence it provides a bijection between the current item and item 48 (E. Deutsch, private communication, December 20, 2006). Deutsch points out that this involution is an unpublished generalization of his note in *Discrete Math.* **204** (1999), 163–166.

50. Let $G = G(x)$ be the (ordinary) generating function for Schröder paths from $(0,0)$ to $(2n,0)$ with no double rises. Hence from item 49 we have $G(x) = (C(x) - 1)/x$. Now every Schröder path with no level steps on the x-axis and no double rises is either empty or has the form (using the notation of the previous item) $UADB$; here either $A = \emptyset$ or $A = LB$, where B is a Schröder path with no double rises. Hence if $F(x)$ is the generating function for Schröder paths from $(0,0)$ to $(2n,0)$ with no level steps on the x-axis and no double rises, then

$$F(x) = 1 + x(1 + xG(x))F(x).$$

It follows easily that $F(x) = C(x)$ (E. Deutsch, private communication, December 20, 2006). Is there a simple bijective proof?

51. There is an obvious bijection with the two-colored Motzkin paths of item 40: replace the step $(1,1)$ with 1, $(1,-1)$ with -1, red $(1,0)$ with 0, and blue $(1,0)$ with 0^*. This item is due to E. Deutsch, private communication, January 20, 2007. Deutsch also notes a linear algebraic interpretation: let $A = (A_{ij})_{i,j \geq 1}$ be the tridiagonal matrix with $A_{ii} = 2$ and $A_{i,i+1} = A_{i,i-1} = 1$. Then $C_n = (A^{n-1})_{11}$.

52. Replace each step $(1,1)$ or $(0,1)$ with the step $(1,1)$, and replace each step $(1,0)$ with $(1,-1)$. We obtain a bijection with the paths of item 28.

53. Given such a path, prepend an up step U and append a down step D. Each maximal segment below ground level in the elevated path has the form $D^k U^k$ for some $k \geq 1$. Replace it with $(UD)^k$. This gives a bijection with the Dyck paths of item 25 (D. Callan, private communication, June 13, 2012).

54. The bijection with Dyck paths has the same description as in item 53, that is, given such a path, each maximal segment below

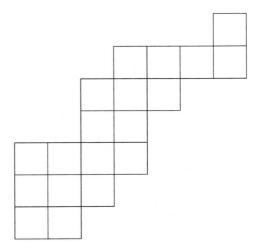

Figure 3.3. A parallelogram polyomino.

ground level has the form $D^k U^k$ for some $k \geq 1$. Replace it with $(UD)^k$.

55. See Section 3 of D. Callan, math.CO/0604471.

56. See R. K. Guy, *J. Integer Sequences* **3** (2000), article 00.1.6, and R. K. Guy, C. Krattenthaler, and B. Sagan, *Ars Combinatorica* **34** (1992), 3–15.

57. The region bounded by the two paths is called a *parallelogram polyomino*. It is an array of unit squares, say with k columns $C_1, \ldots, C_k$. Let a_i be the number of squares in column C_i, for $1 \leq i \leq k$, and let b_i be the number of rows in common to C_i and C_{i+1}, for $1 \leq i \leq k-1$. Define a sequence σ of 1's and -1's as follows (where exponentiation denotes repetition):

$$\sigma = 1^{a_1} (-1)^{a_1 - b_1 + 1} 1^{a_2 - b_1 + 1} (-1)^{a_2 - b_2 + 1} 1^{a_3 - b_2 + 1}$$
$$\cdots 1^{a_k - b_{k-1} + 1} (-1)^{a_k}.$$

This sets up a bijection with item 77 (ballot sequences). For the parallelogram polyomino of Figure 3.3 we have $(a_1, \ldots, a_7) = (3,3,4,4,2,1,2)$ and $(b_1, \ldots, b_6) = (3,2,3,2,1,1)$. Hence (writing $-$ for -1)

$$\sigma = 111 - 1 - -111 - -11 - - -1 - -1 - 11 - -.$$

The enumeration of parallelogram polyominoes is due to J. Levine, *Scripta Math.* **24** (1959), 335–338, and later G. Pólya, *J. Combinatorial Theory* **6** (1969), 102–105. See also L. W. Shapiro, *Discrete Math.* **14** (1976), 83–90; W.-J. Woan, L. W. Shapiro, and D. G. Rogers, *Am. Math. Month.* **104** (1997), 926–931; G. Louchard, *Random Struct. Algor.* **11** (1997), 151–178; and R. A. Sulanke, *J. Differ. Equ. Appl.* **5** (1999), 155–176. For more information on the fascinating topic of polyomino enumeration, see M. Delest and G. Viennot, *Theor. Comput. Sci.* **34** (1984), 169–206, and X. G. Viennot, in *Séries formelles et combinatoire algébrique* (P. Leroux and C. Reutenauer, eds.), Publications de Laboratoire de Combinatoire er d'Informatique Mathématique, vol. 11, Université du Québec à Montréal, 1992, pp. 399–420.

58. Regarding a path as a sequence of steps, remove the first and last steps from the two paths in item 57. This variation of item 57 was suggested by L. W. Shapiro, private communication, 1998.

59. Fix a vertex v. Starting clockwise from v, at each vertex write 1 if encountering a chord for the first time and -1 otherwise. This gives a bijection with item 77. This result is apparently due to A. Errera, *Mém. Acad. Roy. Belgique Coll. 8° (2)* **11** (1931), 26pp. See also J. Riordan, *Math. Comp.* **29** (1975), 215–222, and S. Dulucq and J.-G. Penaud, *Discrete Math.* **17** (1993), 89–105.

60. A simple combinatorial proof was given by L. W. Shapiro, *J. Combinatorial Theory* **20** (1976), 375–376. Shapiro observes that this result is a combinatorial manifestation of the identity

$$\sum_{k \geq 0} \binom{n}{2k} 2^{n-2k} C_k = C_{n+1},$$

due to J. Touchard, in *Proc. Int. Math. Congress, Toronto (1924)*, vol. 1, 1928, p. 465.

61. Cut the circle in item 59 between two fixed vertices and "straighten out." An interpretation of noncrossing matchings as "quantum states of an assembly of $2n$ electrons, with total spin zero" is given by H.N.V. Temperley and E. H. Lieb, *Proc. Roy. Soc. London A* **322** (1971), 251–280. See page 263.

62, 63. These results are due to I. M. Gelfand, M. I. Graev, and A. Postnikov, in *The Arnold-Gelfand Mathematical Seminars*, Birkhäuser, Boston, 1997, pp. 205–221 (§6). For item 62, note that there is always an arc from the leftmost to the rightmost vertex. When this arc is removed, we obtain two smaller trees satisfying the conditions of the problem. This leads to an easy bijection with item 4. The trees of item 62 are called *noncrossing alternating trees*.

An equivalent way of stating the above bijection is as follows. Let T be a noncrossing alternating tree on the vertex set $1, 2, \ldots, n+1$ (in that order from left to right). Suppose that vertex i has r_i neighbors that are larger than i. Let u_i be the word in the alphabet $\{1, -1\}$ consisting of r_i 1's followed by a -1. Let $u(T) = u_1 u_2 \cdots u_{n+1}$. Then u is a bijection between the objects counted by items 62 and 77. It was shown by M. Schlosser that exactly the same definition of u gives a bijection between items 63 and 77! The proof, however, is considerably more difficult than in the case of item 62. (A more complicated bijection was given earlier by C. Krattenthaler.)

For further information on trees satisfying conditions (a), (b), and (d) (called *alternating trees*), see [65, Exercise 5.41].

64. Replace the left-hand endpoint of each arc with a 1 and the right-hand endpoint with a -1. We claim that this gives a bijection with the ballot sequences of item 77. First note that if we do the same construction for the noncrossing matchings of item 61, then it is very easy to see that we get a bijection with item 77. Hence we will give a bijection from item 61 to 64 with the additional property that the locations of the left endpoints and right endpoints of the arcs are preserved. (Of course any bijection between items 61 and 64 would suffice to prove the present item; we are showing a stronger result.)

Let M be a noncrossing matching on $2n$ points. Suppose that we are given the set S of left endpoints of the arcs of M. We can recover M by scanning the elements of S from right to left, and attaching each element i to the *leftmost* available point to its right. In other words, draw an arc from i to the first point to the right of i that does not belong to S and to which no arc has already been attached. If we change this algorithm by attaching each element of S to the *rightmost* available point to its right, then it can be checked that we obtain a nonnesting matching and that we have defined a bijection from item 61 to 64.

Bijections between noncrossing and nonnesting matchings go back to M. De Saintes-Catherine, Thèse du 3me cycle, Université de Bordeaux I, 1983. See also A. de Médicis and X. G. Viennot, *Adv. in Appl. Math.* **15** (1994), 262–304, and A. Kasraoui and J. Zeng, *Electr. J. Comb.* **13** (2006), R33. For further information on crossings and nestings of matchings, see W. Chen, E. Deng,

R. Du, R. Stanley, and C. Yan, *Trans. Amer. Math. Soc.* **359** (2007), 1555–1575, and the references given there.

65. Let $f : \mathbb{P} \to \mathbb{P}$ be any function satisfying $f(i) \leq i$. Given a ballot sequence $\alpha = (a_1, \ldots, a_{2n})$ as in item 77, define the corresponding *f-matching* M_α as follows. Scan the 1's in α from right to left. Initially all the 1's and -1's in α are unpaired. When we encounter $a_i = 1$ in α, let j be the number of unpaired -1's to its right, and draw an arc from a_i to the $f(j)$th -1 to its right (thus pairing a_i with this -1). Continue until we have paired a_1, after which all terms of α will be paired, thus yielding the matching M_α. By construction, the number of f-matchings of $[2n]$ is C_n. This gives infinitely many combinatorial interpretations of C_n, but of course most of these will be of no special interest. If $f(i) = 1$ for all i, then we obtain the noncrossing matchings of item 61. If $f(i) = i$ for all i, then we obtain the nonnesting matchings of item 64. If $f(i) = \lfloor i/2 \rfloor$ for all i, then we obtain the matchings of the present item. Thus these matchings are in a sense "halfway between" noncrossing and nonnesting matchings.

66. Reading the points from left to right, replace each isolated point and each point that is the left endpoint of an arc with 1, and replace each point that is the right endpoint of an arc with -1. We obtain a bijection with item 98.

67. Label the points $1, 2, \ldots, n$ from left to right. Given a noncrossing partition of $[n]$ as in item 159, draw an arc from the first element of each block to the other elements of that block, yielding a bijection with the current item. This result is related to research on network testing done by N. Kube, private communication from F. Ruskey, November 9, 2004.

68. Given a binary tree with n vertices as in item 4, add a new root with a left edge connected to the old root. Label the $n + 1$ vertices by $1, 2, \ldots, n + 1$ in preorder. For each right edge, draw an arc from its bottom vertex to the top vertex of the first left edge encountered on the path to the root. An example is shown in Figure 3.4. On the left is a binary tree with $n = 5$ vertices; in the middle is the augmented tree with $n + 1$ vertices with the preorder labeling; and on the right is the corresponding set of arcs. This item is due to D. Callan, private communication, March 23, 2004.

69. If we label the vertices $1, \ldots, n + 1$ from right to left, then we obtain a lopsided representation of the trees of item 16 (D. Callan, private communication, May 28, 2010).

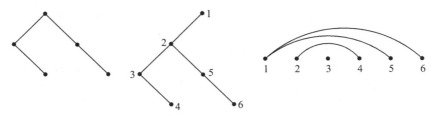

Figure 3.4. The bijection for Problem 68.

70. Replace the left-hand endpoint of each arc by U (an up step), and replace the right-hand endpoint by D (a down step). Replace each isolated point in even position by U, and in odd position by D. Prepend U and append D to get a bijection with the Dyck paths of item 25. This item is due to D. Callan, private communication, September 13, and September 14, 2007.

71. Read the vertices from left to right. Replace an isolated vertex by $1, -1$, the left endpoint of an arc by 1, and the right endpoint of an arc with -1 to obtain a bijection with the ballot sequences of item 77. This item is due to E. Deutsch, private communication, February 27, 2007. D. Bevan pointed out (private communication, August 25, 2010) that if we replace each isolated vertex with two adjacent vertices connected by an arc, then we obtain a bijection with item 61.

72. See S. Elizalde, "Statistics on Pattern-Avoiding Permutations," PhD thesis, MIT, June 2004 (Proposition 3.5.3(1)).

73. Three proofs are given by H. Niederhausen, *Electr. J. Comb.* **9** (2002), #R33.

74. Let P be a parallelogram polyomino of the type being counted. Linearly order the maximal vertical line segments on the boundary of P according to the level of their bottommost step. Replace such a line segment appearing on the right-hand (respectively, left-hand) path of the boundary of P by a 1 (respectively, -1), but omit the final line segment (which will always be on the left). For instance, for the first parallelogram polyomino shown in the statement of the problem, we get the sequence $(1, 1, -1, -1, 1, -1)$. This sets up a bijection with item 77. This result is due to E. Deutsch, S. Elizalde, and A. Reifegerste, private communication, April 2003.

75. If we consider the number of chord diagrams of item 59 containing a fixed horizontal chord, then we obtain the standard quadratic recurrence for Catalan numbers. An elegant "bijectivization" of this

argument is the following. Fix a vertex v. Given a nonintersecting chord diagram with a distinguished horizontal chord K, rotate the chords so that the left-hand endpoint of K is v. This gives a bijection with item 59. Another way to say this (suggested by R. Chapman) is that there are n different chord slopes, each occuring the same number of times, and hence C_n times.

76. Kepler towers were created by X. Viennot, who gave a bijection with the Dyck paths of item 25. Viennot's bijection was written up by D. E. Knuth, Three Catalan bijections, Institut Mittag-Leffler preprint series, 2005 spring, #04; www.ml.kva.se/ preprints/0405s. The portion of this paper devoted to Kepler walls is also available at www-cs-faculty.stanford.edu/ ~knuth/programs/viennot.w.

78. Consider a lattice path P of the type of item 24. Let a_{i-1} be the area above the x-axis between $x = i - 1$ and $x = i$, and below P. This sets up a bijection as desired.

79. Subtract $i - 1$ from a_i and append a one at the beginning to get item 78. This result is closely related to Problem A8(c). If we replace the alphabet $1, 2, \ldots, 2(n - 1)$ with the alphabet $\overline{n-1}, n - 1, \overline{n-2}, n - 2, \ldots, \overline{1}, 1$ (in that order) and write the new sequence $b_1, b_2, \ldots, b_{n-1}$ in reverse order in a column, then we obtain the arrays of R. King, in *Lecture Notes in Physics*, vol. 50, Springer-Verlag, Berlin/Heidelberg/New York, 1975, pp. 490–499 (see also S. Sundaram, *J. Combinatorial Theory Ser. A* **53** (1990), 209–238 (Def. 1.1)) that index the weights of the $(n - 1)$-st fundamental representation of $Sp(2(n - 1), \mathbb{C})$.

80. Let $b_i = a_i - a_{i+1} + 1$. Replace a_i with one 1 followed by b_i -1's for $1 \le i \le n$ (with $a_{n+1} = 0$) to get item 77.

81. Take the first differences of the sequences in item 80.

82. Do a depth-first search through a plane tree with $n + 1$ vertices as in item 6. When a vertex is encountered for the first time, record one less than its number of children, except that the last vertex is ignored. This gives a bijection with item 6.

83. These sequences are just the inversion tables (as defined in [64, §1.3]) of the 321-avoiding permutations of item 115. For a proof, see S. C. Billey, W. Jockusch, and R. Stanley, *J. Alg. Combinatorics* **2** (1993), 345–374 (Thm. 2.1). (The previous reference deals with the *code* $c(w)$ of a permutation w rather than the inversion table $I(w)$. They are related by $c(w) = I(w^{-1})$. Since w is

321-avoiding if and only if w^{-1} is 321-avoiding, it makes no difference whether we work with the code or with the inversion table.)

84. If we replace a_i by $n - a_i$, then the resulting sequences are just the inversion tables of 213-avoiding permutations w (i.e., there does not exist $i < j < k$ such that $w_j < w_i < w_k$). Such permutations are in obvious bijection with the 312-avoiding permutations of item 116. For further aspects of this exercise, see [65, Exercise 6.32].

85. Given a sequence $a_1, \ldots, a_n$ of the type being counted, define recursively a binary tree $T(a_1, \ldots, a_n)$ as follows. Set $T(\emptyset) = \emptyset$. If $n > 0$, then let the left subtree of the root of $T(a_1, \ldots, a_n)$ be $T(a_1, a_2, \ldots, a_{n-a_n})$ and the right subtree of the root be $T(a_{n-a_n+1}, a_{n-a_n+2}, \ldots, a_{n-1})$. This sets up a bijection with item 4. Alternatively, the sequences $a_n - 1, a_{n-1} - 1, \ldots, a_1 - 1$ are just the inversion tables of the 312-avoiding permutations of item 116. Let us also note that the sequences $a_1, a_2, \ldots, a_n$ are precisely the sequences $\tau(u)$, $u \in \mathfrak{S}_n$, of [65, Exercise 5.49(d)].

86. Add 1 to the terms of the sequences of item 82. Alternatively, if $(b_1, \ldots, b_{n-1})$ is a sequence of item 78, then let $(a_1, \ldots, a_n) = (n + 1 - b_{n-1}, b_{n-1} - b_{n-2}, \ldots, b_2 - b_1)$.

87. Let L be the lattice path from $(0,0)$ to (n,n) with steps $(0,1)$ and $(1,0)$ whose successive horizontal steps are at heights $a_1, a_2, \ldots, a_n$. (In particular, the first step must be $(0,1)$.) There will be exactly one horizontal step ending at a point (i,i). Reflect the portion of L from $(0,0)$ to (i,i) about the line $y = x$. We obtain a bijection with the lattice paths counted by item 24. This result is due to W. Moreira, private communication from M. Aguiar, October 2005.

88. The positions of each i that occurs form the blocks of the noncrossing partitions of item 159. In fact, the sequences of this item are the "restricted growth functions" of [64, Exercise 1.106] corresponding to noncrossing partitions. See also M. Klazar, *Europ. J. Combinatorics* **17** (1996), 53–68. This item was suggested by D. Callan, private communication, January 16, 2011.

89. Obvious bijection with item 79. The sequences being counted are also the positions of the down steps in the Dyck paths of length $2n$ in item 25, and the second rows of the standard Young tableaux of item 168.

90. These sequences are just the complements (within the set $[2n]$) of those of item 89. They are also the positions of the up steps in the Dyck paths of length $2n$ in item 25, and the first rows of the standard Young tableaux of item 168.

91. This result was obtained in collaboration with Y. Hu in 2011. See Y. Hu, *J. Int. Seq.* **17** (2004), Article 14.1.6. To get a bijection with the Dyck paths D of item 25, let $a_i - 1$ be the number of down steps before the ith peak, and let b_i be the number of up steps in D before the ith peak.

92. See J. H. van Lint, Combinatorial Theory Seminar, Eindhoven University of Technology, Lecture Notes in Mathematics, no. 382, Springer-Verlag, Berlin/Heidelberg/ New York, 1974 (pp. 22 and 26–27).

93. Fix a root vertex v of a convex $(n + 2)$-gon, and label the other vertices $0, 1, \ldots, n$ in clockwise order from v. If the vertex i is connected to v, then set $a_i = 1$. If in a triangle with vertices $i < j < k$ we have already computed a_i and a_k, then let $a_j = a_i + a_k$. This sets up a bijection with item 1. This item was provided by J. Stevens, private communication, June 10, 2009.

94. Partial sums of the sequences in item 77. The sequences of this item appear explicitly in E. P. Wigner, *Ann. Math.* **62** (1955), 548–564.

95. This interpretation was conjectured by M. Albert, N. Ruškuc, and V. Vatter, reported by Vatter at `http://mathoverflow.net/ questions/131585`. At this website a bijection with Dyck paths was given by C. Stump, later followed by the paper *J. Int. Seq.* **17** (2014), Article 14.7.1.

96. See Theorem 9 of S. Kitaev and J. Remmel, *Discrete Mathematics & Theoretical Computer Science*, Proceedings of the 22nd International Conference on Formal Power Series and Algebraic Combinatorics (2010), 821–832. The authors use the term *restricted ascent sequence* for the sequences being enumerated. Their Theorem 9 involves collaboration with A. Claesson, M. Dukes, and H. H. Gudmundsson.

97. These sequences are the depths of the leaves of the complete binary trees of item 5, read in preorder. This item is due to F. Hivert, private communication, October 12, 2011.

98. In item 28, replace an up step with 1 and a down step with -1.

99. T. Gobet, `arXiv:1405.1408` (Proposition 2.10) constructs a bijection with noncrossing partitions (item 159) in connection with the Bruhat order on noncrossing partitions.

100. Let $B = (b_1, b_2, \ldots, b_{2n})$ be a ballot sequence. Define a subsequence S of $(a_1, a_2, \ldots, a_{2n-2})$ as follows. Let a_{2i-1} belong to S if the ith -1 in B is followed by a 1. Let a_{2i} belong to S if the $(i+1)$st 1 in B is preceded by a -1. For instance, if $B = (1, 1, 1, -1, 1, 1, -1, 1, -1, -1, -1)$, then $S = (a_3, a_5, a_6, a_{10})$. The correspondence $B \mapsto S$ is a bijection between item 77 and the present item.

101. Suppose that the reverse sequence $b_1 \cdots b_{2n-2} = a_{2n-2} \cdots a_1$ begins with k -1's. Remove these -1's, and for each $1 \le i \le k$ remove the rightmost b_j for which $b_{k+1} + b_{k+2} + \cdots + b_j = i$. This yields a sequence of $k+1$ ballot sequences as given by item 77. Place a 1 at the beginning and -1 at the end of each of these ballot sequences and concatenate, yielding a bijection with item 77. This result (stated in terms of lattice paths) is due to D. Callan, private communication, February 26, 2004.

Example. Let $a_1 \cdots a_{14} = + - - + - - + + + + - + - -$ (writing $+$ for 1 and $-$ for -1), so $b_1 \cdots b_{14} = - - - + - + + + + - - + - - +$. Remove b_1, b_2, b_5, b_{14}, yielding the ballot sequences $+-$, $+++--+--$, and $\emptyset$. We end up with the ballot sequence $+ + - - + + + + - - + - - + + -$.

102. Consider the pairs of lattice paths of item 57 and the lines L_i defined by $x + y = i$, $1 \le i \le n$. Let S denote the set of all lattice squares contained between the two paths. The line L_i will pass through the interior of some b_i elements of S. Set $a_i = \pm b_i$ as follows:

- $a_1 = b_1 = 1$
- $a_i a_i > 0$ if $b_i \ne b_{i-1}$
- Suppose that $b_i = b_{i-1}$. Then $a_i = a_{i-1}$ if the top lattice square in S that L_i passes through lies above the top lattice square in S that L_{i-1} passes through, and otherwise $a_i = -a_{i-1}$.

This sets up a bijection with item 57.

103. In the tree T of item 23, label the root by 0 and the two children of the root by 0 and 1. Then label the remaining vertices

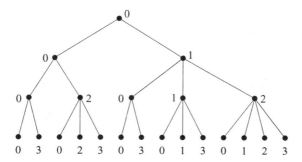

Figure 3.5. The tree for item 103.

recursively as follows. Suppose that the vertex v has height n and is labeled by j. Suppose also that the siblings of v with labels less than j are labeled $t_1, \ldots, t_i$. It follows that v has $i + 2$ children, which we label $t_1, \ldots, t_i, j, n$. See Figure 3.5 for the labeling up to height 3. As in the second solution to item 23, a saturated chain from the root to a vertex at level $n - 1$ is thus labeled by a sequence $(a_1, a_2, \ldots, a_n)$. It can be seen that this sets up a bijection between level $n - 1$ and the sequences we are trying to count. The proof then follows from item 23. This exercise is due to Z. Sunik, *Electr. J. Comb.* **10** (2003), N5. Sunik also points out that the number of elements labeled j at level n is equal to $C_j C_{n-j}$.

104. Given a plane tree with n edges, traverse the edges in preorder and record for each edge except the last the degree (number of children) of the vertex terminating the edge. It is easy to check that this procedure sets up a bijection with item 6. This result is due to D. Callan, private communication, November 3, 2004.

105. Let T be a plane tree with $n + 1$ vertices labeled $1, 2, \ldots, n + 1$ in preorder. Do a depth first search through T and write down the vertices in the order they are visited (including repetitions). This establishes a bijection with item 6. The sequences of this exercise appear implicitly in E. P. Wigner, *Ann. Math.* **62** (1955), 548–564, viz., as a contribution $X_{a_1 a_2} X_{a_2 a_3} \cdots X_{a_{2n-1} a_{2n}} X_{a_{2n} a_1}$ to the $(1, 1)$-entry of the matrix X^{2n}. Item 94 is related.

106. The sequences $1, 1 + a_n, 1 + a_n + a_{n-1}, \ldots, 1 + a_n + a_{n-1} + \cdots + a_2$ coincide with those of item 78. See R. Stanley, *J. Combinatorial Theory* **14** (1973), 209–214 (Theorem 1).

107. Partially order the set $P_n = \{(i,j) : 1 \le i < j \le n\}$ componentwise. Then the sets $\{(i_1,j_1),\ldots,(i_k,j_k)\}$ are just the antichains of P_n and hence are equinumerous with the order ideals of P_n (see the end of [64, §3.1]). But P_n is isomorphic to the poset $\mathrm{Int}(\boldsymbol{n-1})$ of item 178, so the proof follows from this item.

This result is implicit in the paper A. Reifegerste, *Eur. J. Combin.* **24** (2003), 759–776. She observes that if $(i_1 \cdots i_k, j_1 \cdots j_k)$ is a pair being counted, then there is a unique 321-avoiding permutation $w \in \mathfrak{S}_n$ whose *excedance set* $E_w = \{i : w(i) > i\}$ is $\{i_1,\ldots,i_k\}$ and such that $w(i_m) = j_m$ for all m. Conversely, every 321-avoiding $w \in \mathfrak{S}_n$ gives rise to a pair being counted. Thus the proof follows from item 115. Note that if we subtract 1 from each j_k, then we obtain a bijection with pairs of sequences $1 \le i_1 < i_2 < \cdots < i_k \le n-1$ and $1 \le h_1 < h_2 < \cdots < h_k \le n-1$ such that $i_r \le j_r$ for all r. This variant was suggested by E. Deutsch, private communication, November 4, 2007.

D. Callan points out (private communication, January 16, 2011) that this item is also implicit in R. Simion, *J. Combinatorial Theory Ser. A* **66** (1994), 270–301 (REMARK 0.2 on page 275). Namely, increment the i's by 1 and prepend a 1 to get her f-vector, and decrement the j's by 1 and append an n to get her ℓ'-vector.

108. Immediate from the generating function identity

$$C(x) = \frac{1}{1 - xC(x)} = 1 + xC(x) + x^2C(x)^2 + \cdots.$$

This result is due to E. Deutsch, private communication, April 8, 2005.

109. Let $\alpha = (\alpha_1,\ldots,\alpha_k)$, $\beta = (\beta_1,\ldots,\beta_k)$. Define a Dyck path by going up α_1 steps, then down β_1 steps, then up α_2 steps, then down β_2 steps, etc. This gives a bijection with (item 25), due to A. Reifegerste, $\mathtt{arXiv:math/0212247}$ (Cor. 3.8).

110. If $a = a_1a_2\cdots a_k$ is a word in the alphabet $[n-1]$, then let $w(a) = s_{a_1}s_{a_2}\cdots s_{a_k} \in \mathfrak{S}_n$, where s_i denotes the adjacent transposition $(i,i+1)$. Then $w(a) = w(b)$ if $a \sim b$; and the permutations $w(a)$, as a ranges over a set of representatives for the classes B being counted, are just those enumerated by item 115. This statement follows from

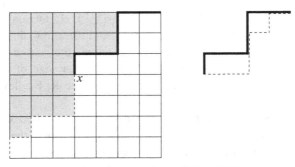

Figure 3.6. A bijection between items 179 and 111.

S. C. Billey, W. Jockusch, and R. Stanley, *J. Algebr. Comb.* **2** (1993), 345–374 (Theorem 2.1).

111. Regard a partition whose diagram fits in an $(n-1) \times (n-1)$ square as an order ideal of the poset $(n-1) \times (n-1)$ in an obvious way. Then the partitions being counted correspond to the order ideals of item 179. Bijections with other Catalan families were given by D. E. Knuth and A. Postnikov. Postnikov's bijection is the following. Let λ be a partition whose diagram is contained in an $(n-1) \times (n-1)$ square S. Let x be the lower right corner of the Durfee square of λ. Let L_1 be the lattice path from the upper right corner of S to x that follows the boundary of λ. Similarly, let L_2 be the lattice path from the lower left corner of S to x that follows the boundary of λ. Reflect L_2 about the main diagonal of S. The paths L_1 and the reflection of L_2 form a pair of paths as in item 58. Figure 3.6 illustrates this bijection for $n = 8$ and $\lambda = (5,5,3,3,3,1)$. The path L_1 is shown in dark lines and L_2 and its reflection in dashed lines.

112. Let $C_\lambda = \cdots c_{-2} c_{-1} c_0 c_1 c_2 \cdots$ be the code of the partition λ, as defined in [65, Exercise 7.59], where the first 1 in the sequence is c_{1-m} such that m is the number of 0's to the right of this 1. Let $S_\lambda = \{i : c_i = 1\}$. For instance, if $\lambda = (3,1,1)$ then $S_\lambda = \mathbb{N} - \{1,2,5\}$. It is easy to see (using [65, Exercise 7.59(b)]) that λ is an n-core and an $(n+1)$-core if and only if S_λ is a set counted by item 201, and the proof follows. This cute result is due to J. Anderson, *Discrete Math.* **248** (2002), 237–243. Anderson obtains the more general result that if m and n are relatively prime, then the number of partitions λ that are both

m-cores and n-cores is $\frac{1}{m+n}\binom{m+m}{m}$. J. Olsson and D. Stanton, *Aequat. Math.* **74** (2007), 90–110, show that in addition the largest $|\lambda|$ for which λ is an m-core and n-core is given by $(m^2 - 1)(n^2 - 1)/24$. For the average size of a partition that is both an n-core and $(n + 1)$-core, see R. Stanley and F. Zanello, arXiv:1313.4352.

113. Remove the first occurrence of each number. What remains is a permutation w of $[n]$ that uniquely determines the original sequence. These permutations are precisely the ones in item 116. There is also an obvious bijection between the sequences being counted and the nonintersecting arcs of item 61.

114. Replace each odd number by 1 and even number by -1 to get a bijection with ballot sequences (item 77).

115. The reference [65, Cor. 7.23.11] shows that the RSK algorithm [65, §7.11] establishes a bijection with item 169. See also D. E. Knuth [34, p. 64].

The earliest explicit enumeration of 321-avoiding permutations is due to P. A. MacMahon [48, §97]. MacMahon states his result not in terms of pattern avoidance, but rather in terms of permutations that are a union of two decreasing sequences. MacMahon's result was rediscovered by J. M. Hammersley, in *Proc. Sixth Berkeley Symposium on Mathematical Statistics and Probability*, vol. 1, University of California Press, Berkeley/Los Angeles, 1972, pp. 345–394. In equation (15.9) he states the result without proof, saying "and this can be proved in general." Proofs were later given by D. E. Knuth [34, §5.1.4] and D. Rotem, *Inf. Proc. Letters* **4** (1975/76), 58–61. Knuth's proof was combinatorial. Another direct combinatorial proof, based on an idea of Goodman, de la Harpe, and Jones, appears in S. C. Billey, W. Jockusch, and R. Stanley, *J. Alg. Combinatorics* **2** (1993), 345–374 (after the proof of Theorem 2.1). A sketch of this proof goes as follows. Given the 321-avoiding permutation $w = a_1 a_2 \cdots a_n$, define $c_i = \#\{j : j > i, w_j < w_i\}$. Let $\{j_1,\ldots,j_\ell\}_< = \{j : c_j > 0\}$. Define a lattice path from $(0,0)$ to (n,n) as follows. Walk horizontally from $(0,0)$ to $(c_{j_1} + j_1 - 1, 0)$, then vertically to $(c_{j_1} + j_1 - 1, j_1)$, then horizontally to $(c_{j_2} + j_2 - 1, j_1)$, then vertically to $(c_{j_2} + j_2 - 1, j_2)$, etc. The last part of the path is a vertical line from $(c_{j_\ell} + j_\ell - 1, j_{\ell-1})$ to $(c_{j_\ell} + j_\ell - 1, j_\ell)$, then (if needed) a horizontal line to $(c_{j_\ell} + j_\ell - 1, n)$, and finally a vertical line to (n,n). This establishes a bijection with item 24.

For an elegant bijection with item 116, see R. Simion and F. W. Schmidt, *Europ. J. Combinatorics* **6** (1985), 383–406 (Prop. 19). Two other bijections with item 116 appear in D. Richards, *Ars Combinatoria* **25** (1988), 83–86, and J. West, *Discrete Math.* **146** (1995) 247–262 (Thm. 2.8). See also [64, §1.5].

116. There is an obvious bijection between 312-avoiding and 231-avoiding permutations, viz., $a_1 a_2 \ldots a_n \mapsto n + 1 - a_n, \ldots, n + 1 - a_2, n + 1 - a_1$. It is easily seen that the 231-avoiding permutations are the same as those of item 119, as first observed by D. E. Knuth [33, Exercise 2.2.1.5]. The enumeration *via* Catalan numbers appears in [33, Exercise 2.2.1.4]. References to bijections with item 115 are given in the solution to item 115.

For a straightforward proof, note that a permutation $u, 1, v$ of $[n + 1]$ is 312-avoiding if and only if every term of u is less than every term of v, and both u and v are 312-avoiding. This observation immediately yields the fundamental recurrence (1.1).

For the problem of counting permutations in $\mathfrak{S}_n$ according to the number of subsequences with the pattern 132 (equivalently, 213, 231, or 312), see M. Bóna, in Conference Proceedings, vol. 1, Formal Power Series and Algebraic Combinatorics, July 14–July 18, 1997, Universität Wien, pp. 107–118.

117. This result appears on p. 796 of D. M. Jackson, *Trans. Amer. Math. Soc.* **299** (1987), 785–801, but probably goes much further back. For a direct bijective proof, it is not hard to show that the involutions counted here are the same as those in item 121.

118. A coding of planar maps due to R. Cori, *Astérisque* **27** (1975), 169 pp., when restricted to plane trees, sets up a bijection with item 6. We can also set up a bijection with noncrossing partitions π (item 159) by letting the cycles of u be the blocks of π with elements written in decreasing order. See S. Dulucq and R. Simion, *J. Algebraic Comb.* **8** (1998), 169–191.

119. When an element a_i is put on the stack, record a 1. When it is taken off, record a -1. This sets up a bijection with item 77. This result is due to D. E. Knuth [33, Exercise 2.2.1.4]. The permutations being counted are just the 231-avoiding permutations, which are in obvious bijection with the 312-avoiding permutations of item 116 (see Knuth, ibid., Exercise 2.2.1.5).

120. Same set as item 115, as first observed by R. Tarjan, *J. Assoc. Comput. Mach.* **19** (1972), 341–346 (the case $m = 2$ of Lemma 2). The concept of queue sorting is due to Knuth [33, chap. 2.2.1].

121. Obvious bijection with item 61.

122. See I. M. Gessel and C. Reutenauer, *J. Comb. Theory Ser. A* **64** (1993), 189–215 (Thm. 9.4 and discussion following).

123. This result is due to O. Guibert and S. Linusson, in *Conference Proceedings*, vol. 2, *Formal Power Series and Algebraic Combinatorics*, July 14–July 18, 1997, Universität Wien, pp. 243–252. For the "limiting shape" of the permutations of this item, see T. Dokos and I. Pak.

   NOTE. The number of alternating Baxter permutations of $1, 2, \ldots, 2n + \delta$ ($\delta = 0$ or 1) is $C_n C_{n+\delta}$, as first shown by R. Cori, S. Dulucq, and G. Viennot, *J. Comb. Theory Ser. A* **43** (1986), 1–22, and later with a direct bijective proof by S. Dulucq and O. Guibert, *Discrete Math.* **180** (1998), 143–156. For the "limiting shape" of alternating Baxter permutations, see Remark 6.8 of Dokos and I. Pak, *Online J. Analytic Comb.* **9** (2014).

124. This is the same set as item 115. See Theorem 2.1 of the reference given in item 115 to S. C. Billey et al. For a generalization to other Coxeter groups, see J. R. Stembridge, *J. Alg. Combinatorics* **5** (1996), 353–385.

125. These are just the 132-avoiding permutations $w_1 \cdots w_n$ of $[n]$ (i.e., there does not exist $i < j < k$ such that $w_i < w_k < w_j$), which are in obvious bijection with the 312-avoiding permutations of item 116. This result is an immediate consequence of the following results: (i) I. G. Macdonald, *Notes on Schubert Polynomials*, Publications du LACIM, vol. 6, Univ. du Québec à Montréal, 1991, (4.7) and its converse stated on p. 46 (due to A. Lascoux and M. P. Schützenberger); (ii) ibid., equation (6.11) (due to Macdonald); (iii) item 116 of this monograph; and (iv) the easy characterization of dominant permutations (as defined in Macdonald, ibid.) as 132-avoiding permutations. For a simpler proof of the crucial (6.11) of Macdonald, see S. Fomin and R. Stanley, *Advances in Math.* **103** (1994), 196–207 (Lemma 2.3).

126. The point is that the permutations 321, 312, and 231 are themselves indecomposable. It follows that if w is any one of 321, 312, 231, then the number $f(n)$ of indecomposable w-avoiding permutations

in $\mathfrak{S}_n$ satisfies

$$s_n(w) = \sum_{k=1}^{n} f(k) s_{n-k}(w),$$

where $s_m(w)$ denotes the total number of permutations in $\mathfrak{S}_n$ avoiding w. Since $s_n(w) = C_n$, we get, just as in the solution to [64, Exercise 1.128(a)], that

$$\sum_{n \geq 1} f(n) x^n = 1 - \frac{1}{C(x)}$$

$$= xC(x),$$

and the proof follows. For some bijective proofs, see Section 4 of A. Claesson and S. Kitaev, *Sém. Lotharingien de Combinatoire* **60** (2008), Article B60d (electronic).

127. For $w = 213$, simply prepend a 1 to a 213-avoiding permutation of $2, 3, \ldots, n+1$. For $w = 132$, simply append a 4 to a 132-avoiding permutation of $1, 2, \ldots, n$.

128. See J. Françon and G. Viennot, *Discrete Math.* **28** (1979), 21–35.

129. Three proofs of this result are discussed by D. Gewurz and F. Merola, *Europ. J. Combinatorics* **27** (2006), 990–994. In particular, each of the sequences $a_1, \ldots, a_n$ and $b_1, \ldots, b_n$ uniquely determines the other. The sequences $a_1, \ldots, a_n$ are those of item 78, while the sequences $b_1, \ldots, b_n$ are the 231-avoiding permutations of $2, 3, \ldots, n+1$ (equivalent to item 116).

130. This result is due to N. Reading, *Trans. Amer. Math. Soc.* **359** (2007), 5931–5958. In Example 6.3 he describes the following bijection with the noncrossing partitions of item 159. The sequences $a_1 a_2 \cdots a_p$ being counted define distinct permutations $w = u_{a_1} u_{a_2} \cdots u_{a_p} \in \mathfrak{S}_n$. Define an equivalence relation $\sim$ on $[n]$ to be the transitive and reflexive closure of $w(i) \sim w(i+1)$ if $w(i) > w(i+1)$. The equivalence classes then form a noncrossing partition of $[n]$. For instance, if $w = 143652$, then the noncrossing partition is 1-34-256.

In the special case $u_i = s_i$, there is a nice direct bijection between the sequences $a_1 a_2 \cdots a_p$ and lattice paths L that never rise above $y = x$ (item 24). Label the lattice squares by their height above the x-axis (beginning with height 1). Read the labels of the squares below the line $y = x$ and above the lattice path L by first reading from bottom to top the leftmost diagonal, then the next leftmost diagonal, etc. This procedure sets up a bijection between item 24 and the sequences $a_1 a_2 \cdots a_p$. The figure below shows a lattice path L with the labeling of the lattice squares. It follows that the sequence corresponding to L is 123456123413.

131. Replace the first occurrence of i with 1 and the second occurrence with -1 to get the ballot sequences of item 77. This item is due to D. Callan, private communication, September 1, 2007.

132. Replace the first occurrence of i with 1 and the second occurrence with -1, and then remove the first 1 and last -1 to get the ballot sequences of item 77. This item is due to C. S. Barnes, PhD thesis, Harvard University, 2009.

133. C. O. Oakley and R. J. Wisner showed (*Amer. Math. Monthly* **64** (1957), 143–154) that hexaflexagons can be represented by pats and deduced that they are counted by Catalan numbers. A simple bijection with complete binary trees (item 5) was given by D. Callan, arXiv:1005.5736 (§4).

134. Deleting the first two entries (necessarily a 1 and a 2) and taking the positions of the first appearance of $3, \ldots, n+1$ in the resulting permutation is a bijection to item 79. If we drop the condition that the first appearances of $1, 2, \ldots, n+1$ occur in order, then the resulting sequences are enumerated by R. Graham and N. Zang, *J. Combinatorial Theory, Ser. A* **115** (2008), 293–303. This item is due to D. Callan, private communication, September 1, 2007. It is also easy to see that the sequences being enumerated coincide with those of item 132.

135. Let M be a noncrossing matching on $2n$ vertices (item 61). Color the vertices black and white alternating from left to right,

Figure 3.7. A noncrossing matching.

beginning with a black vertex. Thus, every arc connects a white vertex to a black one. If an arc connects the ith black vertex to the jth white vertex, then set $w(i) = j$. This sets up a bijection with the permutations being counted. For instance, if M is given by Figure 3.7 then $w = 623154$. This result was suggested by E. Y.-P. Deng, private communication, June 2010.

136. The bijection between genus 0 permutations and noncrossing partitions in the solution to item 118 shows that we are counting noncrossing partitions of $[n + 1]$ for which 1 is a singleton. These are clearly in bijection with noncrossing partitions of $[2, n + 1]$.

This item and the next three are due to E. Deutsch, private communication, May 31, 2010.

137. Now we are counting noncrossing partitions for which 1 and 2 are in the same block.

138. Now 1 and 3 are in the same block, and 2 is a singleton.

139. Now $\{1, n + 2\}$ is a doubleton block. Of course there are infinitely many more items of this nature.

140. See N. A. Loehr, *Europ. J. Combinatorics* **26** (2005), 83–93. This contrived-looking interpretation of C_n is actually closely related to Exercise 6.25(i) and the (q, t)-Catalan numbers of Garsia and Haiman. See Problem A44 for additional information on these polynomials.

141. See N. A. Loehr, ibid.

142. The restriction of such w to the last n terms gives a bijection with 123-avoiding permutations in $\mathfrak{S}_n$ (equivalent to item 115). Namely, symmetric permutations with last n entries $1, \ldots, n$ in some 123-avoiding order are clearly 123-avoiding. Conversely, if there were a number $i > n + 1$ among the last n entries, then we would have the increasing subsequence $2n + 2 - i, n + 1, i$. This result (with a different proof, in a more general context) appears in E. S. Egge, *Annals of Comb.* **11** (2007), 405–434 (pp. 407, 412).

143. Let $S_v(n)$ denote the set of all permutations $w \in \mathfrak{S}_n$ avoiding the permutation $v \in \mathfrak{S}_k$, and let fp(w) denote the number of fixed points of w. S. Elizalde, "Statistics on Pattern-Avoiding Permutations," PhD thesis, MIT, June 2004 (Section 5.6) and *Electron. J. Combin.* **18** (2011), P29 ($q = 1$ case of equation (2)), has shown that

$$\sum_{n \geq 0} \sum_{w \in S_{321}(n)} x^{\text{fp}(w)} t^n = \frac{2}{1 + 2t(1-x) + \sqrt{1-4t}}. \tag{3.1}$$

Applying $\frac{d}{dx}$ and setting $x = 1$ gives $C(t) - 1$, and the proof follows. This paper also contains a bijection from 321-avoiding permutations w in $\mathfrak{S}_n$ to Dyck paths P of length $2n$. This bijection takes fixed points of w into peaks of height one of P. Now use item 38 to get a bijective proof of the present item. This item is due to E. Deutsch, private communication, September 6, 2007. For further statistical properties of fixed points of 123-avoiding permutations (and the 132-avoiding permutations of the next item), see S. Miner and I. Pak, *J. Advances in Applied Math.* **55** (2014), 86–130 (§7).

Note that the average number of fixed points of a 321-avoiding permutation in $\mathfrak{S}_n$ is one, the same as the average number of fixed points of all permutations in $\mathfrak{S}_n$. What other "interesting" classes of permutations have this property?

144. See A. Robertson, D. Saracino, and D. Zeilberger, *Annals of Comb.* **6** (2003), 427–444. For a bijection with fixed points of 321-avoiding permutations (item 143), see S. Elizalde and I. Pak, *J. Combinatorial Theory, Ser. A* **105** (2004), 207–219.

145. Replace an excedance of w with a 1 and a nonexcedance with a -1, except for the nonexcedance $w(2n + 1)$ at the end of w. This sets up a bijection with item 77. There is also a close connection with item 74. If P is a parallelogram polyomino of the type counted by item 74, then place P in a $(2n + 1) \times (2n + 1)$ square M. Put a 1 in each square immediately to the right of the bottom step in each maximal vertical line on the boundary, except for the rightmost such vertical line. Put a 0 in the remaining squares of M. This sets up a bijection between item 74 and the permutation matrices corresponding to the permutations counted by the present exercise. An example is given by the figure below, where the corresponding permutation is 4512736. This result is due to

E. Deutsch, S. Elizalde, and A. Reifegerste, private communication, April 2003.

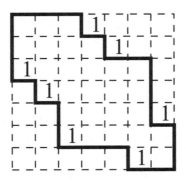

146. Let $a_1 a_2 \cdots a_{2n}$ be a permutation being counted, and associate with it the array

$$\begin{array}{ccccc} a_2 & a_4 & a_6 & \cdots & a_{2n} \\ a_1 & a_3 & a_5 & \cdots & a_{2n-1} \end{array}.$$

This sets up a bijection with item 168, standard Young tableaux of shape (n,n). This result is due to E. Deutsch and A. Reifegerste, private communication, June 4, 2003. Deutsch and Reifegerste also point out that the permutations being counted have an alternative description as those 321-avoiding permutations in $\mathfrak{S}_{2n}$ with the maximum number of descents (or equivalently, excedances), namely n.

147. This result can be proved using the machinery of R. Stanley, *J. Combinatorial Theory, Ser. A* **114** (2007), 436–460. Presumably there is a simpler combinatorial proof. Note that it suffices to assume that n is even, since $a_1, a_2, \ldots, a_{2m}$ is reverse-alternating and 321-avoiding if and only if the same is true for $a_1, a_2, \ldots, a_{2m-1}, 2m+1, a_{2m}$. (By similar reasoning, item 146 is reduced to the case where n is odd.)

NOTE. The results of items 146 and 147 can easily be carried over to alternating 123-avoiding and reverse alternating 123-avoiding permutations by considering the reverse $a_n \cdots a_1$ of the permutation $a_1 \cdots a_n$.

148. The second entry has to be 1. Delete it to obtain a 321-avoiding permutation of $\{2, 3, \ldots, n+1\}$, enumerated by item 115. This item is due to E. Deutsch, private communication, June 18, 2009.

149, 150. Let $f(m)$ denote the number of alternating 132-avoiding permutations in $\mathfrak{S}_m$, and let $w = w_1 w_2 \cdots w_m \in \mathfrak{S}_m$ be such a permutation. Then $w_{2i+1} = m$ for some i. Moreover, $w_1 w_2 \cdots w_{2i}$ is an alternating 132-avoiding permutation of $m - 2i, m - 2i + 1, \ldots, m - 1$, while $w_{2i+2}, w_{2i+3}, \ldots, w_m$ is a reverse alternating permutation of $1, 2, \ldots, m - 2i - 1$. By induction we obtain the formula

$$f(2n) = C_0 C_{n-1} + C_1 C_{n-2} + \cdots + C_{n-1} C_0 = C_n.$$

The argument for $f(2n + 1)$, as well as for reverse alternating 132-avoiding permutations, is analogous. These results are due to T. Mansour, *Annals of Comb.* **7** (2003), 201–227. (Theorem 2.2). Mansour obtains similar results for 132-avoiding permutations that are alternating or reverse alternating except at the first step, and he gives numerous generalizations and extensions. For further work in this area, see J. B. Lewis, *Electronic J. Combinatorics* **16(1)** (2009), #N7.

151. By [65, Cor. 7.13.6] (applied to permutation matrices), [65, Thm. 7.23.17] (in the case $i = 1$), and [65, Exercise 7.28(a)] (in the case where A is a symmetric permutation matrix of trace 0), the RSK algorithm sets up a bijection between 321-avoiding fixed-point-free involutions in $\mathfrak{S}_{2n}$ and standard Young tableaux of shape (n, n). Now use item 168. There are also numerous ways to give a more direct bijection.

152. As in item 151, the RSK-algorithm sets up a bijection between 321-avoiding involutions in $\mathfrak{S}_{2n-1}$ with one fixed point and standard Young tableaux of shape $(n, n - 1)$; and again use item 168.

153. In the solution to item 119 it was mentioned that a permutation is stack-sortable if and only if it is 231-avoiding. Hence a permutation in $\mathfrak{S}_{2n}$ can be sorted into the order $2n, 2n - 1, \ldots, 1$ on a stack if and only if it is 213-avoiding. Given a 213-avoiding fixed-point-free involution in $\mathfrak{S}_{2n}$, sort it in reverse order on a stack. When an element is put on the stack, record a 1, and when it is taken off, record a -1 (as in the solution to item 77). Then we obtain exactly the sequences $a_1, a_2, \ldots, a_n, -a_n, \ldots, -a_2, -a_1$, where $a_1, a_2, \ldots, a_n$ is a ballot sequence (item 77), and the proof follows. Moreover, E. Deutsch (private communication, May 2001) has constructed a bijection with the Dyck paths of item 25.

154. Similar to item 153.

155. Obvious bijection with item 60.

156. If we arrange the elements of S in increasing order, then we obtain the sequences of item 89. This result is equivalent to one of B. Hopkins and M. A. Jones, in *The Mathematics of Preference, Choice and Order*, Springer, Berlin, 2009, pp. 273–285. For some further developments, see L. J. Billera, L. Levine, and K. Mészáros, *Proc. Amer. Math, Soc.*, to appear; arXiv:1306.6744.

157. Let L be a lattice path as in item 24. Let $(0,0) = v_0, v_1, \ldots, v_k = (n,n)$ be the successive points at which L intersects the diagonal $y = x$. Let L' be the path obtained by reflecting about $y = x$ the portions of L between each v_{2i-1} and v_{2i}. The horizontal steps of L' then correspond to the moves of the first player, while the vertical steps correspond to the moves of the second player.

   This result and solution are due to Lou Shapiro, private communication, May 13, 2005. Shapiro stated the result in terms of the game of Parcheesi, but since many readers may be unfamiliar with this game we have given a more mundane formulation.

158. Since $n + 1$ and n are relatively prime, each equivalence class has exactly $2n + 1$ elements. Hence the number of classes is $\frac{1}{2n+1}\binom{2n+1}{n} = C_n$. This fact is the basis for our direct combinatorial proof (Section 1.6) that there are C_n ballot sequences of length $2n$.

159. Let T be a plane tree with $n + 1$ vertices (item 6). Label the nonroot vertices $1, 2, \ldots, n$ in preorder. The sets of children of nonleaf vertices gives a bijection with noncrossing partitions.

   Noncrossing partitions first arose in the work of H. W. Becker, *Math. Mag.* **22** (1948–49), 23–26, in the form of *planar rhyme schemes*, i.e., rhyme schemes with no crossings in the *Puttenham diagram*, defined by G. Puttenham, *The Arte of English Poesie*, London, 1589 (pp. 86–88). Further results on noncrossing partitions are given by H. Prodinger, *Discrete Math.* **46** (1983), 205–206; N. Dershowitz and S. Zaks, *Discrete Math.* **62** (1986), 215–218; R. Simion and D. Ullman, *Discrete Math.* **98** (1991), 193–206; P. H. Edelman and R. Simion, *Discrete Math.* **126** (1994), 107–119; R. Simion, *J. Combinatorial Theory Ser. A* **65** (1994); R. Speicher, *Math. Ann.* **298** (1994), 611–628; A. Nica and R. Speicher, *J. Algebraic Combinatorics*, **6** (1997), 141–160;

R. Stanley, *Electron. J. Combinatorics* **4**, R20 (1997), 14 pp.
See also [64, Exercise 3.158] (the case $k = 1$) and [65, Exercise 5.35].

160. These partitions are clearly the same as the noncrossing partitions of item 159. This description of noncrossing partitions is due to R. Steinberg, private communication.

161. Obvious bijection with item 159. (Vertical lines are in the same block if they are connected by a horizontal line.) Murasaki diagrams were used in *The Tale of Genji*, written by Lady Murasaki (or Murasaki Shikibu), c. 978–c. 1014 or 1025, to represent the 52 partitions of a five-element set. The noncrossing Murasaki diagrams correspond exactly to the noncrossing partitions. The statement that noncrossing Murasaki diagrams are enumerated by Catalan numbers seems to have been observed first by H. W. Gould, who pointed it out to M. Gardner, leading to its mention in M. Gardner, *Scientific American* **234** (June 1976), pp. 120–125, and bibliography on p. 132; reprinted (with an Addendum) in Chapter 20 of *Time Travel and Other Mathematical Bewilderments*, W. H. Freeman, New York, 1988. Murasaki diagrams were not actually used by Lady Murasaki herself. It wasn't until the Wasan period of old Japanese mathematics, from the late 1600s well into the 1700s, that the Wasanists started attaching the Murasaki diagrams (which were actually incense diagrams) to illustrated editions of *The Tale of Genji*.

162. This result was proved by M. Klazar, *Europ. J. Combinatorics* **17** (1996), 53–68 (p. 56), using generating function techniques.

163. See R. C. Mullin and R. G. Stanton, *Pacific J. Math.* **40** (1972), 167–172 (p. 168). They set up a bijection with item 6. They also show that $2n + 1$ is the largest possible value of k for which there exists a noncrossing partition of $[k]$ with $n + 1$ blocks such that no block contains two consecutive integers. A simple bijection with item 1 was given by D. P. Roselle, *Utilitas Math.* **6** (1974), 91–93. The following bijection with item 6 is due to A. Vetta (1997). Label the vertices $1, 2, \ldots, 2n + 1$ of a tree in item 6 in preorder. Define i and j to be in the same block of $\pi \in \Pi_{2n+1}$ if j is a right child of i.

164. Let P_n denote the poset of intervals with at least two elements of the chain n, ordered by inclusion. Let $\mathcal{A}_n$ denote the set of antichains of P_n. By the last paragraph of [64, §3.1], $\#\mathcal{A}_n$ is equal to the number of order ideals of P_n. But P_n is isomorphic

to the poset $\text{Int}(\boldsymbol{n}-\boldsymbol{1})$ of *all* (nonempty) intervals of $\boldsymbol{n}-\boldsymbol{1}$, so by item 178 we have $\#\mathcal{A}_n = C_n$. Given an antichain $A \in \mathcal{A}_n$, define a partition π of $[n]$ by the condition that i and j (with $i < j$) belong to the same block of π if $[i,j] \in A$ (and no other conditions not implied by these). This establishes a bijection between $\mathcal{A}_n$ and the nonnesting partitions of $[n]$. For a further result on nonnesting partitions, see the solution to [65, Exercise 5.44]. The present exercise was obtained in collaboration with A. Postnikov. The concept of nonnesting partitions for any reflection group (with the present case corresponding to the symmetric group $\mathfrak{S}_n$) is due to Postnikov and is further developed in C. A. Athanasiadis, *Electronic J. Combinatorics* **5** (1998), R42.

165. See D. Callan, *Discrete Math.* **309** (2009), 4187–4191 (§2.4).

166. We claim that each equivalence class contains a unique element $(a_1,\ldots,a_n)$ satisfying $a_1 + a_2 + \cdots + a_i \geq i$ for $1 \leq i \leq n$. The proof then follows from item 86. To prove the claim, if $\alpha = (a_1,\ldots,a_n) \in S_n$, then define $\alpha' = (a_1 - 1,\ldots,a_n - 1,-1)$. Note that the entries of α' are greater than or equal to -1 and sum to -1. If $E = \{\alpha_1,\ldots,\alpha_k\}$ is an equivalence class, then it is easy to see that the set $\{\alpha'_1,\ldots,\alpha'_k\}$ consists of all conjugates (or cyclic shifts) that end in -1 of a single word α'_1, say. It follows from [65, Lemma 5.3.7] that there is a unique conjugate (or cyclic shift) β of α'_1 such that all partial sums of β, except for the sum of all the terms, are nonnegative. Since the last component of β is -1, it follows that $\beta = \alpha'_j$ for a unique j. Let $\alpha_j = (a_1,\ldots,a_n)$. Then α_j will be the unique element of E satisfying $a_1 + \cdots + a_i \geq i$, as desired.

A straightforward counting proof of this item appears in S. K. Pun, PhD thesis, Higher Order Derivatives of the Perron Root, Polytechnic Institute of New York University, 1994. For a generalization, see E. Deutsch and I. Gessel, Problem 10525, solution by D. Beckwith, *Amer. Math. Monthly* **105** (1998), 774–775.

167. If $\lambda = (\lambda_1,\ldots,\lambda_{n-1}) \subseteq (n-1,n-2,\ldots,1)$, then the sequences $(1,\lambda_{n-1}+1,\ldots,\lambda_1+1)$ are in bijection with item 78. Note also that the set of Young diagrams contained in $(n-1,n-2,\ldots,1)$, ordered by inclusion (i.e., the interval $[\emptyset,(n-1,n-2,\ldots,0)]$ in Young's lattice, as defined in [64, Example 3.4.4(b)]), is

isomorphic to $J(\mathrm{Int}(\boldsymbol{n}-\mathbf{1}))^*$, thereby setting up a bijection with item 178.

168. Given a standard Young tableau T of shape (n,n), define $a_1 a_2 \cdots a_{2n}$ by $a_i = 1$ if i appears in row 1 of T, while $a_i = -1$ if i appears in row 2. This sets up a bijection with item 77. See also [34, p. 63] and [65, Prop. 7.10.3].

169. See the solution to item 115 (first paragraph) for a bijection with 321-avoiding permutations. An elegant bijection with item 168 appears in [48, vol. 1, §100] (repeated in [34, p. 63]). Namely, given a standard Young tableau T of shape (n,n), let P consist of the part of T containing the entries $1,2,\ldots,n$; while Q consists of the complement in T of P, rotated 180°, with the entry i replaced by $2n+1-i$. See also [65, Cor. 7.23.12].

170. Given a standard Young tableau T of the type being counted, construct a Dyck path of length $2n$ as follows. For each entry $1,2,\ldots,m$ of T, if i appears in row 1 then draw an up step, while if i appears in row 2 then draw a down step. Afterward, draw an up step followed by down steps to the x-axis. This sets up a bijection with item 25 (Dyck paths).

171. The bijection of item 170 yields an *elevated* Dyck path, i.e., a Dyck path of length $2n+2$ which never touches the x-axis except at the beginning and end. Remove the first and last step to get a bijection with item 25.

172. Remove all entries except $3,5,7,\ldots,2n-1$ and shift the remaining entries in the first row one square to the left. Replace $2i+1$ with i. This sets up a bijection with SYT of shape (n,n), so the proof follows from item 168. This result is due to T. Chow, H. Eriksson, and C. K. Fan, *Electr. J. Comb.* **11**(2) (2004–2005), #A3. This paper also shows the more difficult result that the number of SYT of shape (n,n,n) such that adjacent entries have opposite parity is the number $B(n-1)$ of Baxter permutations of length $n-1$ (defined in [65, Exercise 6.55]).

173. Let b_i be the number of entries in row i that are equal to $n-i+1$ (so $b_n = 0$). The sequences $b_n + 1, b_{n-1} + 1, \ldots, b_1 + 1$ obtained in this way are in bijection with item 78.

174. These arrays encode the labeled trees of item 15. The integer r is the depth of the tree, the sequence $a_1, \ldots, a_r$ records the number of edges at each level starting at the bottom, and the sequence

$$\begin{array}{ccc|cc}
2 & 2 & 2 & 1 & 1 \\
2 & 2 & 0 & 0 & 0 \\
1 & 1 & 0 & 0 & 0 \\
1 & 0 & 0 & 0 & 0
\end{array}$$

Figure 3.8. A plane partition and two lattice paths.

$b_1,\ldots,b_{r-1}$ lists the nonleaf vertex labels decremented by 1. For instance, the labeled tree

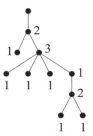

has $r = 5$ and code $\begin{pmatrix} 2 & 1 & 4 & 2 & 1 \\ 1 & 0 & 2 & 1 & \end{pmatrix}$. This item is due to D. Callan, private communication, February 19, 2008.

175. Given the plane partition π, let L be the lattice path from the lower left to upper right that has only 2's above it and no 2's below. Similarly let L' be the lattice path from the lower left to upper right that has only 0's below it and no 0's above. See Figure 3.8 for an example. This pair of lattice paths coincides with those of item 57.

176. This result is equivalent to Proposition 2.1 of S. C. Billey, W. Jockusch, and R. Stanley, *J. Algebraic Combinatorics* **2** (1993), 345–374. See also the last paragraph on page 363 of this reference.

177. Obvious bijection with item 168. This interpretation of Catalan numbers appears in R. Stanley, *Fibonacci Quart.* **13** (1975), 215–232. Note also that if we label the elements of $\mathbf{2} \times \mathbf{n}$ (the direct product of the chains **2** and **n**) analogously to what was illustrated for $n = 3$, then the linear extensions coincide with the permutations of item 114.

178. There is an obvious bijection with order ideals I of $\text{Int}(n)$ that contain every one-element interval of n. But the "upper boundary" of the Hasse diagram of I "looks like" the Dyck paths of item 25. See R. Stanley, ibid., bottom of page 222.

179. This result is equivalent to the $q = 1$ case of G. E. Andrews, *J. Stat. Plan. Inf.* **34** (1993), 19–22 (Corollary 1). For a more explicit statement and some generalizations, see R. G. Donnelly, PhD thesis, University of North Carolina, 1997, and *J. Combinatorial Theory Ser. A* **88** (1999), 217–234. For a bijective proof, see the solution to item 111. A sequence of posets interpolating between the poset $\text{Int}(n-1)$ of item 178 and A_{n-1}, and all having C_n order ideals, was given by D. E. Knuth, private communication, December 9, 1997.

180. Given a sequence $1 \leq a_1 \leq \cdots \leq a_n$ of integers with $a_i \leq i$, define a poset P on the set $\{x_1, \ldots, x_n\}$ by the condition that $x_i < x_j$ if and only if $j + a_{n+1-i} \geq n + 1$. (Equivalently, if Z is the matrix of the zeta function of P, then the 1's in $Z - I$ form the shape of the Young diagram of a partition, rotated 90° clockwise and justified into the upper right-hand corner.) This yields a bijection with item 78. This result is due to R. L. Wine and J. E. Freund, *Ann. Math. Statis.* **28** (1957), 256–259. See also R. A. Dean and G. Keller, *Canad. J. Math.* **20** (1968), 535–554. Such posets are called *semiorders* or *unit interval orders*. For further information, see P. C. Fishburn, *Interval Orders and Interval Graphs*, Wiley-Interscience, New York, 1985; and W. T. Trotter, *Combinatorics and Partially Ordered Sets*, Johns Hopkins University Press, Baltimore/London, 1992 (ch. 8). For the labeled version of this item, see Problem A39.

181. The lattice $J(P)$ of order ideals of the poset P has a natural planar Hasse diagram. There will be two elements covering $\hat{0}$, corresponding to the two minimal elements of P. Draw the Hasse diagram of $J(P)$ so that the rooted minimal element of P goes to the left of $\hat{0}$ (so the other minimal element goes to the right). The "outside boundary" of the Hasse diagram then "looks like" the pair of paths in item 57 (rotated 45° counterclockwise).

182. Such posets are known as *series-parallel interval orders*. They are enumerated by F. Disanto, L. Ferrari, R. Pinzani, and S. Rinaldi, in *Associahedra, Tamari Lattices and Related Structures,*

F. Müller-Hoissen, J. M. Pallo, and J. Stasheff, eds., Progress in Mathematics, vol. 299, 2012, Springer, Basel, pp. 323–338. To obtain a bijection with item 61 (noncrossing matchings), let M be a noncrossing matching on $[2n]$. Define a partial order on the arcs by letting $a < b$ if a lies entirely to the left of b. Compare item 180.

183. Let A be an antichain of the poset of intervals of the chain $n - 1$. The number of such antichains is C_n by item 178, since for any poset there is a simple bijection between its order ideals I and antichains A, viz., A is the set of maximal elements of I. (See [64, (3.1)].) Construct from A a poset P on the points $1, 2, \ldots, n + 1, 1', 2', \ldots, (n + 1)'$ as follows. First, $1 < 2 < \cdots < n + 1$ and $1' < 2' < \cdots < (n + 1)'$. If $[i, j] \in A$, then define $i < (j + 2)'$ and $j < (i + 2)'$. This gives a bijection between items 178 and 183. This result is due to J. Stembridge, private communication, November 22, 2004. Compare item 181.

184. Let $w \in \mathfrak{S}_n$ be 213-avoiding (equivalent to item 116). Let $\varphi(w) = \{(i, j) : i < j, \ w(i) < w(j)\}$. This sets up a bijection with the set of pairs $i <_P j$ for one of the posets P being enumerated. This item (stated differently) is due to S. Sam, private communication, October 23, 2009.

185. A. Claesson and S. Linusson, *Proc. Amer. Math. Soc.* **139** (2011), 435–449, describe on page 445 a bijection with nonnesting matchings (item 64). They also point out that no two of the posets being counted are isomorphic (as unlabeled posets), and that the unlabeled posets appearing here are just the semiorders (i.e., $(\mathbf{3 + 1})$-free and $(\mathbf{2 + 2})$-free posets) on n vertices of item 180.

186. See R. Dewji, I. Dimitrov, A. McCabe, M. Roth, D. Wehlau, and J. Wilson, `arXiv:1110.5880` (Lemma 6.1). This surprising result is related to the faces of the Littlewood-Richardson cone.

187. These relations are called *similarity relations*. See L. W. Shapiro, *Discrete Math.* **14** (1976), 83–90; V. Strehl, *Discrete Math.* **19** (1977), 99–101; D. G. Rogers, *J. Combinatorial Theory Ser. A* **23** (1977), 88–98; J. W. Moon, *Discrete Math.* **26** (1979), 251–260. Moon gives a bijection with item 77. E. Deutsch (private communication) has pointed out an elegant bijection with item 24, viz., the set enclosed by a path and its reflection in the diagonal *is* a similarity relation (as a subset of $[n] \times [n]$). The connectedness

of the columns ensures the last requirement in the definition of a similarity relation. For a connection with the *Catalan simplicial set* of M. Buckley, R. Garner, S. Lack, and R. Street, see arXiv:1309.6120 (Prop. 2.2).

If we think of a similarity relation as a graph on the vertex set $[n]$ with an edge ij if iRj and $i \neq j$, then these graphs are the *unit interval graphs*, that is, graphs whose vertices are some closed intervals of length one on the real line, with two intervals adjacent if they intersect. The vertices are labeled $1, 2, \ldots, n$ in the same order as the left-hand endpoint of the corresponding interval. Unit interval graphs are also the incomparability graphs of the semiorders of item 180.

188. These are just the 321-avoiding permutations $w \in \mathfrak{S}_n$ of item 115. This result appears (without explicitly stating that the answer is C_n) as [64, Exercise 3.185(i)].

189. Call a subset $T \subseteq I_m$ satisfying the condition a *valid m-set*. Write D_m for the number of valid m-sets. Let S be a valid $(n+1)$-set. Let i be the least integer for which $(i, 0) \in S$. (Set $i = n$ if there is no such integer.) The set $J_1 = \{(h, k) \in I_{n+1} : h > i\}$ is a poset isomorphic to I_{n-i}. Choose a valid $(n-i)$-set S_1 in J_1 after identifying it with I_{n-i}. Similarly, $J_2 = \{(h, k) \in I_{n+1} : k > n - i\}$ is isomorphic to I_i. Choose a valid $(n-i)$-set S_2 in J_2 after identifying it with I_i. It can be checked that there is a unique subset T of $I_{n+1} - J_1 - J_2$ for which $S_1 \cup S_2 \cup T$ is a valid $(n+1)$-set. There are D_{n-i} choices for S_1 and D_i choices for S_2, so

$$D_{n+1} = \sum_{i=0}^{n} D_i D_{n-i}.$$

This recurrence, together with the initial condition $D_0 = 1$, shows that $D_n = C_n$.

This result was obtained (with a more geometric description) by J. Stevens, in *Deformations of Surface Singularities*, Bolyai Society Mathematical Studies, vol. 23, Springer, Berlin, 2013, pp. 203–228 (Theorem 3), in connection with versal deformations of cyclic quotient singularities.

190. This result is due to F. Disanto, L. Ferrari, R. Pinzani, and S. Rinaldi, *Elec. Notes Discrete Math.* **34** (2009), 429–433. To obtain a bijection with item 61 (noncrossing matchings), let M be a noncrossing matching on $[2n]$. Define binary relations S and R

on the arcs of M as follows. Let aSb if a lies underneath b, and define aRb if a lies entirely to the left of b. We get all nonisomorphic pairs (S,R) under consideration exactly once, though this takes some work to prove.

191. Obvious bijection with item 178. This interpretation in terms of stacking coins is due to J. Propp. See A. M. Odlyzko and H. S. Wilf, *Amer. Math. Monthly* **95** (1988), 840–843 (Rmk. 1).

192. The total number of n-element multisets on $\mathbb{Z}/(n+1)\mathbb{Z}$ is $\binom{2n}{n}$ (see [64, §1.2]). Call two such multisets M and N *equivalent* if for some $k \in \mathbb{Z}/(n+1)\mathbb{Z}$ we have $M = \{a_1,\ldots,a_n\}$ and $N = \{a_1 + k,\ldots,a_n + k\}$. This defines an equivalence relation in which each equivalence class contains $n+1$ elements, exactly one of which has its elements summing to 0. Hence the number of multisets with elements summing to 0 (or to any other fixed element of $\mathbb{Z}/(n+1)\mathbb{Z}$) is $\frac{1}{n+1}\binom{2n}{n}$. This result appears in R. K. Guy, *Amer. Math. Monthly* **100** (1993), 287–289 (with a more complicated proof due to I. Gessel).

193. Analogous to item 192, using $\frac{1}{n}\binom{2n}{n+1} = C_n$. This problem was suggested by S. Fomin.

194. This result is implicit in the paper G. X. Viennot, *Astérisque* **121–122** (1985), 225–246. Specifically, the bijection used to prove equation (12) of Viennot's paper, when restricted to Dyck words, gives the desired bijection. A simpler bijection follows from the work of J.-G. Penaud, in *Séminaire Lotharingien de Combinatoire*, 22e Session, Université Louis Pasteur, Strasbourg, 1990, pp. 93–130 (Cor. IV-2-8). Yet another proof follows from more general results of J. Bétréma and J.-G. Penaud, *Theoret. Comput. Sci.* **117** (1993), 67–88. For some related problems, see Problem A54 and [65, Exercise 6.46].

195. Let I be an order ideal of the poset Int$(n-1)$ defined in item 178. Associate with I the set R_I of all points $(x_1,\ldots,x_n) \in \mathbb{R}^n$ satisfying $x_1 > \cdots > x_n$ and $x_i - x_j < 1$ if $[i, j-1] \in I$. This sets up a bijection between item 178 and the regions R_I being counted. This result is implicit in R. Stanley, *Proc. Natl. Acad. Sci. U.S.A.* **93** (1996), 2620–2625 (§2), and also appears (as part of more general results) in C. A. Athanasiadis, PhD thesis, MIT, 1996 (Cor. 7.1.3); and A. Postnikov and R. Stanley, *J. Combinatorial Theory Ser. A* **91** (2000), 544–597 (Prop. 7.2).

The set of hyperplanes of this item is known as the *Catalan arrangement*.

196. The solution to item 195 sets up a bijection between order ideals of Int($n-1$) and *all* regions into which the cone $x_1 \geq x_2 \geq \cdots \geq x_n$ is divided by the hyperplanes $x_i - x_j = 1$, for $1 \leq i < j \leq n$. In this bijection, the bounded regions correspond to the order ideals containing all singleton intervals $[i,i]$. It is easy to see that such order ideals are in bijection with *all* order ideals of Int($n-2$). Now use item 178. This result was suggested by S. Fomin. Both this item and item 195 can also be proved using Zaslavsky's theorem (e.g., [64, Thm. 3.11.7]) in the theory of hyperplane arrangements.

197. Let P be a convex $(n+2)$-gon with vertices $v_1, v_2, \ldots, v_{n+2}$ in clockwise order. Let T be a triangulation of T as in item 1, and let a_i be the number of triangles incident to v_i. Then the map $T \mapsto (a_1, \ldots, a_{n+2})$ establishes a bijection with item 1. This remarkable result is due to J. H. Conway and H.S.M. Coxeter, *Math. Gaz.* **57** (1973), 87–94, 175–183 (problems (28) and (29)). The arrays (2.1) are called *frieze patterns*. For a generalization, see C. Bessenrodt, T. Holm, and P. Jørgensen, *J. Combinatorial Theory Ser. A* **123** (2014), 30–42.

198. The objects being counted are known as *Catalan alternative tableaux*. A bijection with complete binary trees (item 5) was given by X. Viennot, *Proc. FPSAC'07, Tianjin, China,* arXiv:0905. 3081. For a connection with the Hopf algebra of binary trees defined by Loday and Ronco, see J.-C. Aval and X. Viennot, *Sém. Lotharingien de Combinatoire* **68** (2010), article B63h.

199. See Definition 15 and Proposition 16 of K. Lee and L. Li, *Electron. J. Combinatorics* **18** (2011), P158. The bijection of this paper was independently found by M. Can and N. Loehr and by A. Woo.

200. In the two-colored Motzkin paths of item 40, number the steps $1, 2, \ldots, n-1$ from left to right. Place the up steps $(1,1)$ in A, the down steps $(1,-1)$ in B, and the red flat steps $(1,0)$ in C. This result is due to D. Callan, private communication, February 26, 2004.

201. Let S_n be the submonoid of $\mathbb{N}$ (under addition) generated by n and $n+1$. Partially order the set $T_n = \mathbb{N} - S_n$ by $i \leq j$ if $j - i \in S_n$. Figure 3.9 illustrates the case $n = 5$. It can be checked that $T_n \cong \text{Int}(n-1)$, as defined in item 178. Moreover,

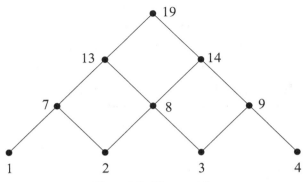

Figure 3.9. The poset T_5.

the subsets S being counted are given by $\mathbb{N} - I$, where I is an order ideal of T_n. The proof follows from item 178. This result is due to M. H. Rosas, private communication, May 29, 2002.

202. Let $S_i - S_{i-1} = \{a_i\}$. Then the sequences $a_1 a_2 \cdots a_n$ coincide with the 312-avoiding permutations of item 116. Communication from R. A. Proctor, June 12, 2012.

203. Let a_i be the multiplicity of $e_i - e_{i+1}$ in the sum. The entire sum is uniquely determined by the sequence $a_1, a_2, \ldots, a_{n-1}$. Moreover, the sequences $0, a_1, a_2, \ldots, a_{n-1}$ that arise in this way coincide with those in item 80. This item is an unpublished result of A. Postnikov and R. Stanley. It is related to the Kostant partition function of type A_{n-1} and the theory of flow polytopes, one reference being W. Baldoni and M. Vergne, *Transform. Groups* **13** (2008), 447–469. Problem A22 is closely related.

204. Let H be a polyomino of the type being counted, say with ℓ rows. Let $a_i + 1$ be the width of the first i rows of H, and let $\alpha_i = a_i - a_{i-1}$ for $1 \le i \le \ell$ (with $a_0 = 0$). Similarly, let $b_i + 1$ be the width of the last i rows of H, and let $\beta_i = b_{\ell-i+1} - b_{\ell-i}$ for $1 \le i \le \ell$ (with $b_0 = 1$). This sets up a bijection with the pairs (α, β) of compositions counted by item 109. This argument is due to A. Reifegerste, as referenced in item 109. By a refinement of this argument she also shows that the number of polyominos of the type being counted with ℓ rows is the Narayana number $N(n, \ell) = \frac{1}{n}\binom{n}{\ell}\binom{n}{\ell-1}$ of Problem A46.

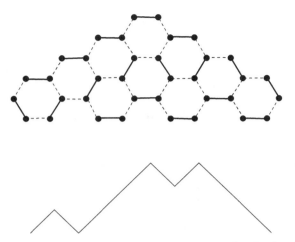

Figure 3.10. A matching on T_4 and the corresponding Dyck path.

Another bijection was provided by E. Deutsch, private communication, June 15, 2001. Namely, given the polynomino H, let $a_1 + 1, \ldots, a_\ell + 1$ be the row lengths and $b_1 + 1, \ldots, b_{\ell-1} + 1$ be the lengths of the overlap between the successive rows. Let D be the Dyck path of length $2n$ with successive peaks at heights $a_1, \ldots, a_\ell$ and successive valleys at heights $b_1, \ldots, b_{\ell-1}$. This sets up a bijection with Dyck paths of length $2n$, as given in item 25. (Compare with the solution to item 57.)

205. There is a simple bijection with the binary trees T of item 4. The root of T corresponds to the rectangle containing the upper right-hand corner of the staircase. Remove this rectangle and we get two smaller staircase tilings, making the bijection obvious. This result is the case $d = 2$ of Theorem 1.1 of H. Thomas, *Order* **19** (2002), 327–342.

206. Concatenate the nonhorizontal edges of the matching and adjoin an up step at the beginning and down step at the end to get a Dyck path as in item 25. See Figure 3.10 for an example. This result is due to S. J. Cyvin and I. Gutman, *Kekulé Structures in Benzenoid Hydrocarbons*, Lecture Notes in Chemistry **46**, Springer-Verlag, Berlin, 1988. The proof given here appears in the interesting survey by T. Došlić, *Croatica Chemica Acta* **78**(2) (2005), 251–259 (Prop. 5.2).

207. See F. T. Leighton and M. Newman, *Proc. Amer. Math. Soc.* **79** (1980), 177–180, and L. W. Shapiro, *Proc. Amer. Math. Soc.* **90** (1984), 488–496.

208. Let

$$(1,1,\ldots,1,-n) = \sum_{i=1}^{n-1} (a_i(e_i - e_{i+1}) + b_i(e_i - e_{n+1})) + a_n(e_n - e_{n+1})$$

as in item 203. Set $m_{ii} = a_i$ and $m_{in} = b_i$. This uniquely determines the matrix M and sets up a bijection with item 203. This result is due to A. Postnikov and R. Stanley (unpublished).

209. See R. Johansson and S. Linusson, *Ann. Combinatorics* **11** (2007), 471–480 (Cor. 3.4). This result is proved by specializing the bijection used to prove Problem A50(z), thereby setting up a bijection with item 48.

210. The monomial $x_1^{a_1} \cdots x_n^{a_n}$ appears in the expansion if and only if the sequence $(a_1,\ldots,a_n)$ is enumerated by item 86. This observation is due to E. Deutsch, private communication, September 27, 2006.

211. If we replace x with 1 and y with -1, then we obtain a bijection with ballot sequences (item 77). The words (terms) appearing in F comprise the *Dyck language*, and the equation $F = 1 + xFyF$ shows that this language is *algebraic* or *context-free*. For further information see [65, §6.6], especially Example 6.6.6, and the related references.

212. Given $w = b_1 b_2 \cdots b_n \in \mathfrak{S}_n$, let $c_i = \min\{b_i, b_{i+1}, \ldots, b_n\}$ and $c(w) = (c_1,\ldots,c_n)$. Then

$$\mathrm{vol}(X_w) = \int_{x_{b_1}=0}^{c_1} \int_{x_{b_2}=x_{b_1}}^{c_2} \cdots \int_{x_{b_n}=x_{b_{n-1}}}^{c_n} dx_1 \, dx_2 \cdots dx_n.$$

Hence $\mathrm{vol}(X_v) = \mathrm{vol}(X_w)$ if $c(v) = c(w)$, and it is not difficult to see that the converse holds. It follows that the number of distinct volumes is the number of distinct sequences $c(w)$, $w \in \mathfrak{S}_n$. These are just the sequences $(c_1,\ldots,c_n) \in \mathbb{P}^n$ satisfying $c_1 \leq c_2 \leq \cdots \leq c_n$ and $c_i \leq i$, and the proof follows from item 78. The problem of finding the number of distinct volumes was raised by Knuth. A solution was given by B. Young, *Comptes Rendus Math.* **348** (2010), 713–716.

213. See L. J. Billera and G. Hetyei, *J. Combinatorial Theory, Ser. A* **89** (2000), 77–104 (Cor. 4).

214. Private communication from M. Aguiar, February 22, 2013. To get a bijection with item 25 (Dyck paths), associate a Dyck path $f(y)$ with an element y of N_1, written with the maximum possible number of x's, as follows. Begin with $f(\emptyset) = \emptyset$. If $y = y_1 + y_2$, then set $f(y) = f(y_1)f(y_2)$ (concatenation of paths). If $y = xz$, then set $f(y) = Uf(z)D$.

4

Additional Problems

In this chapter we give a host of additional problems related to Catalan numbers and their variants and generalizations. Solutions appear in the next chapter. Each problem is given a (subjective) difficulty rating from [1] to [5] using the same scheme as [64] and [65]:

1. routine, straightforward
2. somewhat difficult or tricky
3. difficult
4. horrendously difficult
5. unsolved

Further gradations are indicated by $+$ and $-$. Thus [1−] denotes an utterly trivial problem, and [5−] denotes an unsolved problem that has received little attention and may not be too difficult. A rating of [2+] denotes about the hardest problem that could be reasonably assigned to a class of graduate students. A few students may be capable of solving a [3−] problem, while almost none could solve a [3] in a reasonable period of time. These ratings assume a mathematical background at about the level of a second-year mathematics graduate student.

A1. [2−] The fundamental recurrence $C_{n+1} = \sum_{k=0}^{n} C_k C_{n-k}$ corresponds to "combining" two structures of total size n and doing an additional operation (such as adding a root) to create a structure of size $n + 1$. It may also seem reasonable to put together two structures of total size n, both of which have size less than n, and to omit the additional operation at the end. We then obtain the sequence $D_1, D_2, \ldots$ defined by

$$D_1 = 1, \quad D_n = \sum_{j=1}^{n-1} D_j D_{n-j} \text{ for } n \geq 2.$$

Express D_n in terms of Catalan numbers.

A2. (a) [2+] Let m, n be integers satisfying $1 \le n < m$. Show by a simple bijection that the number of lattice paths from $(1,0)$ to (m,n) with steps $(0,1)$ and $(1,0)$ that intersect the line $y = x$ in at least one point is equal to the number of lattice paths from $(0,1)$ to (m,n) with steps $(0,1)$ and $(1,0)$.

(b) [2−] Deduce that the number of lattice paths from $(0,0)$ to (m,n) with steps $(1,0)$ and $(0,1)$ that intersect the line $y = x$ only at $(0,0)$ is given by $\frac{m-n}{m+n}\binom{m+n}{n}$.

(c) [1+] Show from (b) that the number of lattice paths from $(0,0)$ to (n,n) with steps $(1,0)$ and $(0,1)$ that never rise above the line $y = x$ is given by the Catalan number $C_n = \frac{1}{n+1}\binom{2n}{n}$. This gives a direct combinatorial proof of item 24.[1]

(d) [2−] Show that $B(m,n) := \frac{m-n}{m+n}\binom{m+n}{n}$ is equal to the number of ways $m + n$ voters can vote sequentially in an election for two candidates A and B so that A receives m votes, B receives n votes, and A is always strictly ahead of B. For this reason $B(m,n)$ is called a *ballot number*.

(e) [2] Show that $\sum_{n=0}^{m-1} B(m,n) = C_m$.

A3. (a) [2+] Let X_n be the set of all $\binom{2n}{n}$ lattice paths from $(0,0)$ to (n,n) with steps $(0,1)$ and $(1,0)$. Define the *excedance* of a path $P \in X_n$ to be the number of integers i such that at least one point (i, i') of P lies above the line $y = x$ (i.e., $i' > i$). Show that the number of paths in X_n with excedance j is independent of j.

(b) [1] Deduce that the number of $P \in X_n$ that never rise above the line $y = x$ is given by the Catalan number $C_n = \frac{1}{n+1}\binom{2n}{n}$, another direct proof of item 24.

A4. [2+] Show (bijectively if possible) that the number of lattice paths from $(0,0)$ to $(2n, 2n)$ with steps $(1,0)$ and $(0,1)$ that avoid the points $(2i - 1, 2i - 1)$, $1 \le i \le n$, is equal to the Catalan number C_{2n}.

A5. (a) [2+] A triangle T of a triangulation of a convex $(n+2)$-gon $\mathcal{P}_{n+2}$ is *internal* if no edge of T is an edge of $\mathcal{P}_{n+2}$. Let $t(n)$ denote the total number of internal triangles in all C_n triangulations of $\mathcal{P}_{n+2}$. Show that

$$t(n) = (n+2)C_{n-1} - 2C_n = 2\binom{2n-3}{n-4}.$$

Thus, the probability that a random triangle in a random triangulation of $\mathcal{P}_{n+2}$ is internal is given by

---

[1] Reference to an item always refers to the items of Chapter 2.

$$\frac{t(n)}{nC_n} = \frac{(n-2)(n-3)}{2n(2n-1)}$$

$$= \frac{1}{4} - \frac{9}{8n} + \frac{15}{16n^2} + \frac{15}{32n^3} + \frac{15}{64n^4} + \cdots.$$

(b) [5–] Is there a direct bijective proof that $t(n) = 2\binom{2n-3}{n-4}$?

A6. [3–] Consider the following chess position.

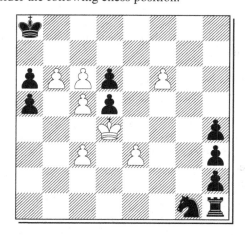

Black is to make 34 consecutive moves, after which White checkmates Black in one move. Black may not move into check, and may not check White (except possibly on his last move). Black and White are *cooperating* to achieve the aim of checkmate. (In chess problem parlance, this problem is called a *serieshelpmate in 34*.) How many different solutions are there?

A7. [?] Explain the significance of the following sequence:

un, dos, tres, quatre, cinc, sis, set, vuit, nou, deu, ...

A8. [2]–[5] Show that the Catalan number $C_n = \frac{1}{n+1}\binom{2n}{n}$ has the algebraic interpretations given below.

(a) Number of two-sided ideals of the algebra of all $(n-1) \times (n-1)$ upper triangular matrices over a field

(b) Dimension of the space of invariants of $SL(2, \mathbb{C})$ acting on the $2n$th tensor power $T^{2n}(V)$ of its "defining" two-dimensional representation V

(c) Dimension of the irreducible representation of the symplectic group $Sp(2(n-1), \mathbb{C})$ (or Lie algebra $\mathfrak{sp}(2(n-1), \mathbb{C})$) with highest weight λ_{n-1}, the $(n-1)$st fundamental weight

(d) Dimension of the primitive intersection homology (say with real coefficients) of the toric variety associated with a (rationally embedded) n-dimensional cube

(e) The generic number of $PGL(2, \mathbb{C})$ equivalence classes of degree n rational maps with a fixed branch set

(f) Number of translation conjugacy classes of degree $n + 1$ monic polynomials in one complex variable, all of whose critical points are fixed

(g) Dimension of the algebra (over a field K) with generators $\epsilon_1, \ldots, \epsilon_{n-1}$ and relations

$$\epsilon_i^2 = \epsilon_i$$

$$\beta \epsilon_i \epsilon_j \epsilon_i = \epsilon_i, \text{ if } |i - j| = 1$$

$$\epsilon_i \epsilon_j = \epsilon_j \epsilon_i, \text{ if } |i - j| \geq 2,$$

where β is a nonzero element of K.

(h) Number of $\oplus$-sign types indexed by A_{n-1}^+ (the set of positive roots of the root system A_{n-1})

(i) Let the symmetric group $\mathfrak{S}_n$ act on the polynomial ring $A = \mathbb{C}[x_1, \ldots, x_n, y_1, \ldots, y_n]$ by $w \cdot f(x_1, \ldots, x_n, y_1, \ldots, y_n) = f(x_{w(1)}, \ldots, x_{w(n)}, y_{w(1)}, \ldots, y_{w(n)})$ for all $w \in \mathfrak{S}_n$. Let I be the ideal generated by all invariants of positive degree, i.e.,

$$I = \langle f \in A : w \cdot f = f \text{ for all } w \in \mathfrak{S}_n, \text{ and } f(0) = 0 \rangle.$$

Then C_n is the dimension of the subspace of A/I affording the sign representation, i.e.,

$$C_n = \dim\{f \in A/I : w \cdot f = (\operatorname{sgn} w)f \text{ for all } w \in \mathfrak{S}_n\}.$$

(j) Degree of the Grassmannian $G(2, n + 2)$ (as a projective variety under the usual Plücker embedding) of 2-dimensional planes in $\mathbb{C}^{n+2}$

(k) Dimension (as a $\mathbb{Q}$-vector space) of the ring $\mathbb{Q}[x_1, \ldots, x_n]/Q_n$, where Q_n denotes the ideal of $\mathbb{Q}[x_1, \ldots, x_n]$ generated by all quasisymmetric functions in the variables $x_1, \ldots, x_n$ with 0 constant term

(l) Multiplicity of the point X_{w_0} in the Schubert variety Ω_w of the flag manifold $GL(n, \mathbb{C})/B$, where $w_0 = n, n - 1, \ldots, 1$ and $w = n, 2, 3, \ldots, n - 2, n - 1, 1$

(m) Conjugacy classes of elements $A \in SL(n, \mathbb{C})$ such that $A^{n+1} = 1$

(n) Number of nonvanishing minors of a generic $n \times n$ upper triangular matrix A (over $\mathbb{C}$, say)

(o) Dimension of the vector space spanned by the products of complementary minors of an $n \times n$ matrix $X = (x_{ij})$ of indeterminates. For instance, when $n = 3$ there are ten such products: $\det(A)$ itself (a product of the empty minor with the full minor) and nine products of 1×1 minors with 2×2 minors, a typical one being $x_{11}(x_{22}x_{33} - x_{23}x_{32})$.

(p) The matrix entry $(A^{2n})_{11}$, where $A = (a_{ij})_{i,j \geq 1}$ in the tridiagonal matrix defined by $a_{i,i-1} = a_{i,i+1} = 1$, and $a_{ij} = 0$ otherwise (and moreover $(A^{2n})_{22} = C_{n+1}$)

(q) Dimension of the space of all complex polynomials $p(x_1, \ldots, x_{2n})$ that are homogeneous of total degree $n(n-1)$ and satisfy

$$p(x_1, \ldots, x_{2n})|_{x_k = \omega x_i = \omega^2 x_j} = 0, \quad 1 \leq i < j < k \leq 2n,$$

where $\omega = e^{2\pi i/3}$.

A9. (a) [3] Let $a_{i,j}(n)$ (respectively, $\bar{a}_{i,j}(n)$) denote the number of walks in n steps from $(0,0)$ to (i,j), with steps $(\pm 1, 0)$ and $(0, \pm 1)$, never touching a point $(-k, 0)$ with $k \geq 0$ (respectively, $k > 0$) once leaving the starting point. Show that

$$a_{0,1}(2n+1) = 4^n C_n$$
$$a_{1,0}(2n+1) = C_{2n+1} \tag{4.1}$$
$$a_{-1,1}(2n) = \frac{1}{2}C_{2n}$$
$$a_{1,1}(2n) = 4^{n-1}C_n + \frac{1}{2}C_{2n}$$
$$\bar{a}_{0,0}(2n) = 2 \cdot 4^n C_n - C_{2n+1}.$$

(b) [3] Show that for $i \geq 1$ and $n \geq i$,

$$a_{-i,i}(2n) = \frac{i}{2n} \frac{\binom{2i}{i}\binom{n+i}{2i}\binom{4n}{2n}}{\binom{2n+2i}{2i}} \tag{4.2}$$

$$a_{i,i}(2n) = a_{-i,-i} + 4^n \frac{i}{n}\binom{2i}{i}\binom{2n}{n-i}. \tag{4.3}$$

(c) [3] Let $b_{i,j}(n)$ (respectively, $\bar{b}_{i,j}(n)$) denote the number of walks in n steps from $(0,0)$ to (i,j), with steps $(\pm 1, \pm 1)$, never touching a point $(-k, 0)$ with $k \geq 0$ (respectively, $k > 0$) once leaving the starting point. Show that

$$b_{1,1}(2n+1) = C_{2n+1}$$

$$b_{-1,1}(2n+1) = 2 \cdot 4^n C_n - C_{2n+1}$$

$$b_{0,2}(2n) = C_{2n}$$

$$b_{2i,0}(2n) = \frac{i}{n}\binom{2i}{i}\binom{2n}{n-i}4^{n-i}, \ i \geq 1 \qquad (4.4)$$

$$\bar{b}_{0,0}(2n) = 4^n C_n. \qquad (4.5)$$

(d) [3–] Let

$$f(n) = \sum_P (-1)^{w(P)},$$

where (i) P ranges over all lattice paths in the plane with $2n$ steps, from $(0,0)$ to $(0,0)$, with steps $(\pm 1,0)$ and $(0,\pm 1)$, and (ii) $w(P)$ denotes the winding number of P with respect to the point $(\frac{1}{2}, \frac{1}{2})$. Show that $f(n) = 4^n C_n$.

A10. [3–] Let $s(T)$ be the minimum number of strokes needed to draw a tree T, where a stroke may not trace over a part of another stroke (except possibly vertices of the tree). Figure 4.1 shows a tree T drawn with the minimum number $s(T) = 3$ of strokes. (Such a drawing is not unique.) Show that for $n \geq 2$ the average value $t(n)$ of $s(T)$ over all C_n binary trees with n vertices (item 4) is given by

$$t(n) = \frac{n}{4} - \frac{1}{8} + \frac{15}{8(2n-1)}.$$

A11. [3] A *rooted planar map* is a planar embedding of an (unlabeled) connected planar graph rooted at a *flag*, i.e., at a triple (v,e,f) where v is a vertex, e is an edge incident to v, and f is a face incident to e. Two rooted planar maps G and H are considered the same if, regarding them as being

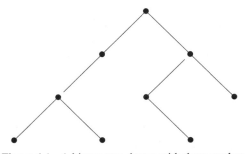

Figure 4.1. A binary tree drawn with three strokes.

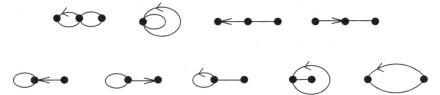

Figure 4.2. The rooted planar maps with two edges.

on the 2-sphere $\mathbb{S}^2$, there is a flag-preserving homeomorphism of $\mathbb{S}^2$ that takes G to H. Equivalently, a rooted planar map may be regarded as a planar embedding of a connected planar graph in which a single edge on the outer face is directed in a counterclockwise direction. (The outer face is the root face, and the tail of the root edge is the root vertex.) Figure 4.2 shows the nine rooted planar maps with two edges.

(a) [3] Show that the number of rooted planar maps with n edges is equal to

$$\frac{2\,(2n)!\,3^n}{n!\,(n+2)!} = \frac{2 \cdot 3^n}{n+2} C_n.$$

(b) [2+] Show that the total number of vertices of all rooted planar maps with n edges is equal to $3^n C_n$.

(c) [3−] Show that the number of pairs (G, T), where G is a rooted planar map with n edges and T is a spanning tree of G, is equal to $C_n C_{n+1}$.

A12. A *k-triangulation* of a convex n-gon C is a maximal collection of diagonals such that there are no $k + 1$ of these diagonals for which any two intersect in their interiors. A 1-triangulation is just an ordinary triangulation, enumerated by the Catalan number C_{n-2} (Theorem 1.5.1(i)). Note that any k-triangulation contains all diagonals between vertices at most distance k apart (where the distance between two vertices u, v is the least number of edges of C we need to traverse in walking from u to v along the boundary of C). We call these nk edges *superfluous*. For example, there are three 2-triangulations of a hexagon, illustrated below (nonsuperfluous edges only).

(a) [3–] Show that all k-triangulations of an n-gon have $k(n - 2k - 1)$ nonsuperfluous edges.
(b) [3] Show that the number $T_k(n)$ of k-triangulations of an n-gon is given by

$$T_k(n) = \det\left[C_{n-i-j}\right]_{i,j=1}^{k}$$

$$= \prod_{1\leq i<j\leq n-2k} \frac{2k+i+j-1}{i+j-1},$$

the latter equality by [64, Thm. 2.7.1] and [65, Exercise 7.101(a)].
(c) [3+] It follows from (b) and [65, Exercise 7.101(a)] that $T_k(n)$ is equal to the number of plane partitions, allowing 0 as a part, of the staircase shape $\delta_{n-2k} = (n - 2k - 1, n - 2k - 2, \ldots, 1)$ and largest part at most k. Give a bijective proof.

A13. [3–] Let P be a Dyck path with $2n$ steps, and let $k_i(P)$ denote the number of up steps in P from level $i - 1$ to level i. Show that

$$\sum_{P,P'} \sum_i k_i(P)k_i(P') = C_{2n} - C_n^2,$$

where the first sum ranges over all pairs (P, P') of Dyck paths with $2n$ steps.

A14. In analogy with our definition of C_n as the number of triangulations of a convex $(n + 2)$-gon, define the *Fuss-Catalan number* $C_{n,k}$ to be the number of dissections (using diagonals that don't intersect in their interiors) of a convex $(kn + 2)$-gon into regions that are $(k + 2)$-gons. Thus $C_{n,1} = C_n$. Many of our combinatorial interpretations and additional properties of Catalan numbers have Fuss-Catalan extensions. We just give a few of the highlights here. Some additional examples, sometimes just for $k = 2$, appear in some of the problems below.
(a) [2+] Show that

$$C_{n,k} = \frac{1}{kn+1}\binom{(k+1)n}{n}.$$

(b) [2–] A *j-ary tree* is a rooted tree such that every vertex has j linearly ordered children, all of which are (possibly empty) j-ary trees. Show that $C_{n,k}$ is the number of $(k + 1)$-ary trees with n vertices.
(c) [2] Show that $C_{n,k}$ is the number of sequences of n k's and $kn - 1$'s such that every partial sum is nonnegative.
(d) [2] Show that $C_{n,k}$ is equal to the number of parenthesizations (or bracketings) of a string of $kn + 1$ letters subject to a nonassociative

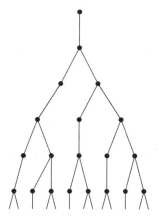

Figure 4.3. The Fibonacci tree.

$(k+1)$-ary operation. For instance, when $k = 2$ and $n = 3$ one such parenthesization is $x(xx(xxx))x$.

(e) [3–] Show that $C_{n,k}$ is the number of *k-divisible noncrossing partitions* of $[kn]$, that is, the number of noncrossing partitions of $[kn]$ for which the size of every block is divisible by k.

(f) [3–] Show that $C_{n,k}$ is equal to the number of 312-avoiding permutations of $[kn]$ such that $1, k+1, 2k+1, \ldots, (n-1)k+1$ is a subsequence (i.e., appear in that order from left to right), and $ik + 1, ik + 2, \ldots, (i+1)k$ are subsequences for $0 \le i \le n - 1$. For example, when $n = 2$ and $k = 3$, the permutations are 123456, 124356, 124536, and 124563.

A15. [2+] The *Fibonacci tree F* is the rooted tree with root v, such that the root has degree one, the child of every vertex of degree one has degree two, and the two children of every vertex of degree two have degrees one and two. Figure 4.3 shows the first six levels of F. Let $f(n)$ be the number of closed walks in F of length $2n$ beginning at v. Show that

$$f(n) = \frac{1}{2n+1}\binom{3n}{n},$$

the number of ternary trees with n vertices.

A16. Let a and b be relatively prime positive integers. The *(a,b)-Catalan number* (not to be confused with the (q,t)-Catalan number of Problem A44) is defined by

$$\text{Cat}(a,b) = \frac{1}{a+b}\binom{a+b}{a}.$$

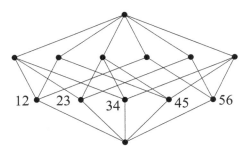

Figure 4.4. The poset $P_{2,3}$ of Problem A18.

(a) [1+] Show that $C_n = \mathrm{Cat}(n+1,n)$. More generally, $\mathrm{Cat}(n,kn+1) = C_{n,k}$, a Fuss-Catalan number (Problem A14).

(b) [3–] Show that $\mathrm{Cat}(a,b)$ is equal to the number of lattice paths, with steps $(1,0)$ and $(0,1)$ from $(0,0)$ to (a,b) never falling below the line from $(0,0)$ to (a,b).

A17. (a) [2+] For $m,n \geq 0$ define

$$S(m,n) = \frac{(2m)!\,(2n)!}{m!\,n!\,(m+n)!},$$

called a *super Catalan number*.[2] Show that $\frac{1}{2}S(m,n)$ is an integer unless $(m,n) = (0,0)$.

(b) [1] Show that $C_n = \frac{1}{2}S(1,n)$ and $\binom{2n}{n} = S(0,n)$.

(c) [5] Find a combinatorial interpretation of $S(m,n)$ or $\frac{1}{2}S(m,n)$.

A18. Let $P_{k,n}$ denote the set of all subsets S of $[kn]$ such that all subsets T of S that are maximal with respect to being a set of consecutive integers have cardinality divisible by k. For instance, when $n \geq 6$ and $k = 3$ we could have $S = \{2,3,4,7,8,9,10,11,12,14,15,16\}$. Partially order $P_{k,n}$ by inclusion. Thus there is a bottom element $\hat{0} = \emptyset$ and top element $\hat{1} = [kn]$. Figure 4.4 shows $P_{2,3}$. Show that

$$\mu(\hat{0},\hat{1}) = (-1)^n \frac{1}{(k-1)n+1}\binom{kn}{n},$$

where μ denotes the Möbius function of $P_{k,n}$. In particular, when $k = 2$ we get $\mu(\hat{0},\hat{1}) = \pm C_n$.

A19. [2+] Let $f(n)$ denote the number of plane ternary trees on the vertex set $[n]$ such that if a vertex i has a child j in the leftmost or center position

---

[2] Sometimes the Schröder numbers s_n of Problem A50 are called super Catalan numbers.

(of the three possible positions for children), then $i < j$. Show that $f(n) = n!C_n$. The four trees for the case $n = 2$ are as follows:

A20. Let $f(n)$ be the number of standard Young tableaux with exactly $2n$ squares, at most four rows, and every row length even. Let $g(n)$ be the number of Young tableaux with exactly $2n$ squares, exactly four rows, and every row length odd. Show that $f(n) - g(n) = C_n$.

A21. Let $t_0, t_1, \ldots$ be indeterminates. If S is a finite subset of $\mathbb{N}$, then set $t^S = \prod_{i \in S} t_i$. For $X = \mathbb{N}$ or $\mathbb{P}$, let $U_n(X)$ denote the set of all n-subsets of X that don't contain two consecutive integers.

    (a) [2+] Show that the following three power series are equal:

       (i) The continued fraction

$$\cfrac{1}{1 - \cfrac{t_0 x}{1 - \cfrac{t_1 x}{1 - \cfrac{t_2 x}{1 - \cdots}}}}$$

    (ii)
$$\frac{\sum_{n \geq 0} (-1)^n \left(\sum_{S \in U_n(\mathbb{P})} t^S \right) x^n}{\sum_{n \geq 0} (-1)^n \left(\sum_{S \in U_n(\mathbb{N})} t^S \right) x^n}$$

    (iii) $\displaystyle\sum_T \prod_{v \in V(T)} t_{\mathrm{ht}(v)}^{\deg(v)} x^{\#V(T)-1}$, where T ranges over all (nonempty)
       plane trees. Moreover, $V(T)$ denotes the vertex set of T, $\mathrm{ht}(v)$
       the height of vertex v (where the root has height 0), and $\deg(v)$
       the degree (number of children) of vertex v.

    The three power series begin

$$1 + t_0 x + (t_0^2 + t_0 t_1)x^2 + (t_0^3 + 2t_0^2 t_1 + t_0 t_1^2 + t_0 t_1 t_2)x^3 + \cdots. \quad (4.6)$$

    (b) [2] Deduce that

$$\lim_{n \to \infty} \frac{\sum_{i \geq 0} (-1)^i \binom{n-i}{i} x^i}{\sum_{i \geq 0} (-1)^i \binom{n+1-i}{i} x^i} = \sum_{k \geq 0} C_k x^k. \quad (4.7)$$

(c) [3] Suppose that $t_0, t_1, \ldots$ are positive integers such that t_0 is odd and $\Delta^k t_n$ is divisible by 2^{k+1} for all $k \geq 1$ and $n \geq 0$, where Δ is the difference operator of [64, §1.9]. For example, $t_i = (2i+1)^2$ satisfies these hypotheses. Let $\sum_{n \geq 0} D_n x^n$ denote the resulting power series (4.6). Show that $v_2(D_n) = v_2(C_n)$, where $v_2(m)$ is the exponent of the largest power of 2 dividing m.

A22. (a) [2+] Start with the monomial $x_{12} x_{23} x_{34} \cdots x_{n,n+1}$, where the variables x_{ij} commute. Continually apply the "reduction rule"

$$x_{ij} x_{jk} \to x_{ik}(x_{ij} + x_{jk}) \tag{4.8}$$

in any order until unable to do so, resulting in a polynomial $P_n(x_{ij})$. Show that $P_n(x_{ij} = 1) = C_n$. (NOTE. The polynomial $P_n(x_{ij})$ itself depends on the order in which the reductions are applied.) For instance, when $n = 3$, one possible sequence of reductions (with the pair of variables being transformed shown in boldface) is given by

$$\mathbf{x_{12} x_{23}} x_{34} \to \mathbf{x_{13}} x_{12} \mathbf{x_{34}} + x_{13} \mathbf{x_{23} x_{34}}$$

$$\to x_{14} x_{13} x_{12} + x_{14} x_{34} x_{12}$$

$$+ x_{14} x_{13} x_{23} + x_{14} \mathbf{x_{34} x_{23}}$$

$$\to x_{14} x_{13} x_{12} + x_{14} x_{34} x_{12}$$

$$+ x_{14} x_{13} x_{23} + x_{14} x_{24} x_{23} + x_{14} x_{24} x_{34}$$

$$= P_3(x_{ij}).$$

(b) [3−] More strongly, replace the rule (4.8) with

$$x_{ij} x_{jk} \to x_{ik}(x_{ij} + x_{jk} - 1),$$

this time ending with a polynomial $Q_n(x_{ij})$. Show that

$$Q_n\left(x_{ij} = \frac{1}{1-x}\right) = \frac{N(n,1) + N(n,2)x + \cdots + N(n,n)x^{n-1}}{(1-x)^n},$$

where $N(n,k)$ is a Narayana number (defined in Problem A46).

(c) [3−] Even more generally, show that

$$Q_n(x_{ij} = t_i) = \sum (-1)^{n-k} t_{i_1} \cdots t_{i_k},$$

where the sum ranges over all pairs $((a_1, a_2, \ldots, a_k), (i_1, i_2, \ldots, i_k)) \in \mathbb{P}^k \times \mathbb{P}^k$ satisfying $1 = a_1 < a_2 < \cdots < a_k \leq n$, $1 \leq i_1 \leq i_2 \leq \cdots \leq i_k$, and $i_j \leq a_j$. (See Problem A50(p).) For instance,

$$Q_3(x_{ij} = t_i) = t_1^3 + t_1^2 t_2 + t_1^2 t_3 + t_1 t_2^2 + t_1 t_2 t_3 - 2t_1^2 - 2t_1 t_2 - t_1 t_3 + t_1.$$

(d) [3+] Now start with the monomial $\prod_{1\leq i<j\leq n+1} x_{ij}$ and apply the reduction rule (4.8) until arriving at a polynomial $R_n(x_{ij})$. Show that

$$R_n(x_{ij}=1)=C_1C_2\cdots C_n.$$

(e) [5–] Generalize (d) in a manner analogous to (b) and (c).

A23. [3–] Let V_r be the operator on (real) polynomials defined by

$$V_r\left(\sum_{i\geq 0}a_iq^i\right)=\sum_{i\geq r}a_iq^i.$$

Define $B_1(q)=-1$, and for $n>1$,

$$B_n(q)=(q-1)B_{n-1}(q)+V_{(n+1)/2}\left(q^{n-1}(1-q)B_{n-1}(1/q)\right).$$

Show that $B_{2n}(1)=B_{2n+1}(1)=(-1)^{n+1}C_n$.

A24. (a) [3+] Let $g(n)$ denote the number of $n\times n$ $\mathbb{N}$-matrices $M=(m_{ij})$ where $m_{ij}=0$ if $j>i+1$, with row and column sum vector $\left(1,3,6,\ldots,\binom{n+1}{2}\right)$. For instance, when $n=2$ there are the two matrices

$$\begin{bmatrix}1&0\\0&3\end{bmatrix}\quad\begin{bmatrix}0&1\\1&2\end{bmatrix},$$

while an example for $n=5$ is

$$\begin{bmatrix}0&1&0&0&0\\0&1&2&0&0\\1&0&3&2&0\\0&0&1&4&5\\0&1&0&4&10\end{bmatrix}.$$

Show that $g(n)=C_1C_2\cdots C_n$.

(b) [2+] Let $f(n)$ be the number of ways to write the vector

$$\left(1,2,3,\ldots,n,-\binom{n+1}{2}\right)\in\mathbb{Z}^{n+1}$$

as a sum of vectors e_i-e_j, $1\leq i<j\leq n+1$, without regard to order, where e_k is the kth unit coordinate vector in $\mathbb{Z}^{n+1}$. For instance, when $n=2$ there are the two ways $(1,2,-3)=(1,0,-1)+2(0,1,-1)=(1,-1,0)+3(0,1,-1)$. Assuming (a), show that $f(n)=C_1C_2\cdots C_n$.

(c) [3–] Let CR_n be the convex polytope of all $n\times n$ doubly stochastic matrices $A=(a_{ij})$ satisfying $a_{ij}=0$ if $i>j+1$. It is easy to see that CR_n is an integral polytope of dimension $\binom{n}{2}$. Assuming (a) or (b), show that the relative volume of CR_n (as defined in [64, p. 497]) is given by

$$v(CR_n) = \frac{C_1 C_2 \cdots C_{n-1}}{\binom{n}{2}!}.$$

A25. [3–] Join $4m + 2$ points on the circumference of a circle with $2m + 1$ nonintersecting chords, as in item 59. Call such a set of chords a *net*. The circle together with the chords forms a map with $2m + 2$ (interior) regions. Color the regions red and blue so that adjacent regions receive different colors. Call the net *even* if an even number of regions are colored red and an even number blue, and *odd* otherwise. The figure below shows an odd net for $m = 2$.

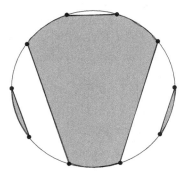

Let $f_e(m)$ (respectively, $f_o(m)$) denote the number of even (respectively, odd) nets on $4m + 2$ points. Show that

$$f_e(m) - f_o(m) = (-1)^{m-1} C_m.$$

A26. [3–] Let $f(n,k)$ be the number of free (unrooted) trees T with $2n$ endpoints (leaves) and k interior vertices satisfying the following properties. The interior vertices of T are unlabeled. No interior vertex of T has degree 2. The exterior vertices are labeled $1, 2, \ldots, n$ and $1', 2', \ldots, n'$. Let e be an edge of T, so that $T - e$ is a forest with two components. Then there is exactly one index $1 \le i \le n$ for which i is one component and i' in the other. There is also exactly one index $1 \le j \le n$ for which j is in one component and $(j + 1)'$ is in the other (where the index $j + 1$ is taken modulo n). Figure 4.5 shows the four such trees with $n = 3$. When $n = 4$ we have $f(4,1) = 1$, $f(4,2) = 8$, $f(4,3) = 12$.

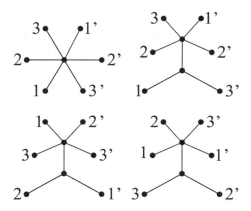

Figure 4.5. The trees of Problem A26 for $n = 3$.

Show that

$$\sum_{k=1}^{n-1}(-1)^{n-1-k}f(n,k) = C_n.$$

A27. [3] Fix $n \geq 1$. Let $f(n)$ be the most number of n-element subsets of $\mathbb{P}$ with the following property. If $a_1 < a_2 < \cdots < a_n$ and $b_1 < b_2 < \cdots < b_n$ are the elements of any two of the subsets, then there exists $i \neq j$ for which $a_i = b_j$. Is it true that $f(n) = C_n$ for all $n \geq 0$?

A28. [5] Define a power series

$$F(x_1,\ldots,x_k) = \sum_{a_1,\ldots,a_k \geq 0} f(a_1,\ldots,a_k)x_1^{a_1}\cdots x_k^{a_k} \in \mathbb{N}[[x_1,\ldots,x_k]]$$

to be $\mathbb{N}$-*rational* if it can be obtained from the series $0, x_1,\ldots,x_k$ by iterating the operations $F + G$ and $F \cdot G$, as well as the operation that takes F to $1/(1 - F)$ provided $F(0,\ldots,0) = 0$. Prove that there exists no such power series satisfying $f(n,n,\ldots,n) = C_n$ for all n.

A29. [3+] Let $\exp\left(E\tan^{-1}(x)\right) = \sum_{n \geq 0} c_n(E)x^n$, where E is an indeterminate. Define

$$A_1(x) = \sum_{n \geq 0} c_n(E)\frac{x^n}{n!}$$

$$A_2(x) = \sqrt{1+x^2}A_1(x)$$

$$A_3(x) = \frac{A_1(x)}{\sqrt{1+x^2}}.$$

For any power series $F(x) = \sum f_n(E)x^n$ let $F'(x) = \sum_{n \geq 1} nf_n(E)x^{n-1}$ and $IF(x) = \sum_{n \geq 0} f_n(E)\frac{x^{n+1}}{n+1}$. For $1 \leq r \leq 3$ define $B_r(x)$ by writing $A_r(x)^2 - A_r'(x) \cdot IA_r(x)$ as a power series in x whose coefficients are polynomials in E, and then replacing each E^j with the Euler number E_j.

(a) Show that the coefficient of $x^{2m-1}/(2m-1)!$ in $B_1(x)$ is C_m.

(b) Show that the coefficient of $x^{2m}/(2m)!$ in $B_2(x)$ is C_{m+1} for $m \geq 2$.

(c) Show that the coefficient of $x^{2m}/(2m)!$ in $B_3(x)$ is C_m.

A30. [2–] The *compositional inverse* $F(x)^{\langle -1 \rangle}$ of a power series $F(x) = a_1 x + a_2 x^2 + \cdots$, where $a_1 \neq 0$, is the unique power series $G(x)$ satisfying $F(G(x)) = G(F(x)) = x$. Find the compositional inverse of the power series $A(x) = xC(x) = x + x^2 + 2x^3 + \cdots$ and $B(x) = C(x) - 1 = x + 2x^2 + 5x^3 + \cdots$.

A31. [2] Find all power series $y = \sum_{n \geq 1} a_n x^n \in \mathbb{C}[[x]]$ such that if $\frac{1}{1-y} = 1 + \sum_{n \geq 1} b_n x^n$, then $\log(1+y) = \sum_{n \geq 1} b_n \frac{x^n}{n}$.

A32. (a) [2+] Show that

$$C(x)^q = \sum_{n \geq 0} \frac{q}{n+q} \binom{2n-1+q}{n} x^n.$$

(b) [2+] Show that

$$\frac{C(x)^q}{\sqrt{1-4x}} = \sum_{n \geq 0} \binom{2n+q}{n} x^n.$$

A33. (a) [2–] Give a generating function proof of the identity

$$\sum_{k=0}^{n} C_{2k} C_{2(n-k)} = 4^n C_n. \tag{4.9}$$

(b) [3–] Give a bijective proof.

A34. (a) [2+] Find all power series $F(t) \in \mathbb{C}[[t]]$ such that if

$$\frac{1 - x + xtF(t)}{1 - x + x^2 t} = \sum_{n \geq 0} f_n(x)t^n,$$

then $f_n(x) \in \mathbb{C}[x]$.

(b) [2+] Find the coefficients of the polynomials $f_n(x)$.

A35. (a) [3–] Let D be a Young diagram of a partition λ, as defined in [64, p. 58]. Given a square s of D, let t be the lowest square in the same column as s, and let u be the rightmost square in the same row as s. Let $f(s)$ be the number of paths from t to u that stay within D, and such that each step is one square to the north or one square to the east. Insert the number $f(s)$ in square s, obtaining an array A. For instance, if $\lambda = (5,4,3,3)$, then A is given by

16	7	2	1	1
6	3	1	1	
3	2	1		
1	1	1		

Let M be the largest square subarray (using consecutive rows and columns) of A containing the upper left-hand corner. Regard M as a matrix. For the above example we have

$$M = \begin{bmatrix} 16 & 7 & 2 \\ 6 & 3 & 1 \\ 3 & 2 & 1 \end{bmatrix}.$$

Show that $\det M = 1$.

(b) [2] Find the unique sequence $a_0, a_1, \ldots$ of real numbers such that for all $n \geq 0$ we have

$$\det \begin{bmatrix} a_0 & a_1 & \cdots & a_n \\ a_1 & a_2 & \cdots & a_{n+1} \\ \cdot & \cdot & & \cdot \\ \cdot & \cdot & & \cdot \\ \cdot & \cdot & & \cdot \\ a_n & a_{n+1} & \cdots & a_{2n} \end{bmatrix} = \det \begin{bmatrix} a_1 & a_2 & \cdots & a_n \\ a_2 & a_3 & \cdots & a_{n+1} \\ \cdot & \cdot & & \cdot \\ \cdot & \cdot & & \cdot \\ \cdot & \cdot & & \cdot \\ a_n & a_{n+1} & \cdots & a_{2n-1} \end{bmatrix} = 1.$$

(When $n = 0$ the second matrix is empty and by convention has determinant one.)

A36. (a) [3–] Let V_n be a real vector space with basis $x_0, x_1, \ldots, x_n$ and scalar product defined by $\langle x_i, x_j \rangle = C_{i+j}$, the $(i+j)$-th Catalan number. It follows from Problem A35(b) that this scalar product is positive definite, and therefore V has an orthonormal basis. Is there an orthonormal basis for V_n whose elements are *integral* linear combinations of the x_i's ?

(b) [3–] Same as (a), except now $\langle x_i, x_j \rangle = C_{i+j+1}$.

(c) [5–] Investigate the same question for the matrices M of Problem A35(a) (so $\langle x_i, x_j \rangle = M_{ij}$) when λ is self-conjugate (so M is symmetric).

A37. (a) [3–] Suppose that real numbers $x_1, x_2, \ldots, x_d$ are chosen uniformly and independently from the interval $[0, 1]$. Show that the probability that the sequence $x_1, x_2, \ldots, x_d$ is convex (i.e., $x_i \le \frac{1}{2}(x_{i-1} + x_{i+1})$ for $2 \le i \le d-1$) is $C_{d-1}/(d-1)!^2$.

(b) [3–] Let $\mathcal{C}_d$ denote the set of all points $(x_1, x_2, \ldots, x_d) \in \mathbb{R}^d$ such that $0 \le x_i \le 1$ and the sequence $x_1, x_2, \ldots, x_d$ is convex. It is easy to see that $\mathcal{C}_d$ is a d-dimensional convex polytope, called the *convexotope*. Show that the vertices of $\mathcal{C}_d$ consist of the points

$$\left(1, \frac{j-1}{j}, \frac{j-2}{j}, \ldots, \frac{1}{j}, 0, 0, \ldots, 0, \frac{1}{k}, \frac{2}{k}, \ldots, 1\right) \tag{4.10}$$

(with at least one 0 coordinate), together with $(1, 1, \ldots, 1)$ (so $\binom{d+1}{2} + 1$ vertices in all). For instance, the vertices of $\mathcal{C}_3$ are $(0,0,0)$, $(0,0,1)$, $(0, \frac{1}{2}, 1)$, $(1,0,0)$, $(1, \frac{1}{2}, 0)$, $(1,0,1)$, $(1,1,1)$.

(c) [3] Show that the Ehrhart quasipolynomial $i(\mathcal{C}_d, n)$ of $\mathcal{C}_d$ (as defined in [64, pp. 494–494]) is given by

$$y_d := \sum_{n \ge 0} i(\mathcal{C}_d, n) x^n$$

$$= \frac{1}{1-x} \left(\sum_{r=1}^{d} \frac{1}{[1][r-1]!} * \frac{1}{[1][d-r]!} \right.$$

$$\left. - \sum_{r=1}^{d-1} \frac{1}{[1][r-1]!} * \frac{1}{[1][d-1-r]!} \right), \tag{4.11}$$

where $[i] = 1 - x^i$, $[i]! = [1][2] \cdots [i]$, and $*$ denotes Hadamard product. For instance,

$$y_1 = \frac{1}{(1-x)^2}$$

$$y_2 = \frac{1+x}{(1-x)^3}$$

$$y_3 = \frac{1+2x+3x^2}{(1-x)^3(1-x^2)}$$

$$y_4 = \frac{1+3x+9x^2+12x^3+11x^4+3x^5+x^6}{(1-x)^2(1-x^2)^2(1-x^3)}$$

$$y_5 = \frac{1+4x+14x^2+34x^3+63x^4+80x^5+87x^6+68x^7+42x^8+20x^9+7x^{10}}{(1-x)(1-x^2)^2(1-x^3)^2(1-x^4)}.$$

Is there a simpler formula than (4.11) for $i(\mathcal{C}_d, n)$ or y_d?

A38. [3] Suppose that $n+1$ points are chosen uniformly and independently from inside a square. Show that the probability that the points are in convex position (i.e., each point is a vertex of the convex hull of all the points) is $(C_n/n!)^2$.

A39. [3–] Let f_n be the number of partial orderings of the set $[n]$ that contain no induced subposets isomorphic to $\mathbf{3} + \mathbf{1}$ or $\mathbf{2} + \mathbf{2}$. (This problem is the labeled analogue of item 180. As mentioned in the solution to this item, such posets are called *semiorders*.) Let $C(x) = 1 + x + 2x^2 + 5x^3 + \cdots$ be the generating function for Catalan numbers. Show that

$$\sum_{n \geq 0} f_n \frac{x^n}{n!} = C(1 - e^{-x}), \qquad (4.12)$$

the composition of $C(x)$ with the series $1 - e^{-x} = x - \frac{1}{2}x^2 + \frac{1}{6}x^3 - \cdots$.

A40. (a) [3–] Let $\mathcal{P}$ denote the convex hull in $\mathbb{R}^{d+1}$ of the origin together with all vectors $e_i - e_j$, where e_i is the ith unit coordinate vector and $i < j$. Thus $\mathcal{P}$ is a d-dimensional convex polytope. Show that the relative volume of $\mathcal{P}$ (as defined in [64, p. 497]) is equal to $C_d/d!$.

(b) [3] Let $i(\mathcal{P}, n)$ denote the Ehrhart polynomial of $\mathcal{P}$. Find a combinatorial interpretation of the coefficients of the i-Eulerian polynomial (in the terminology of [64, p. 473])

$$(1 - x)^{d+1} \sum_{n \geq 0} i(\mathcal{P}, n) x^n.$$

A41. (a) [3–] Define a partial order T_n on the set of all binary bracketings (parenthesizations) of a string of length $n + 1$ as follows. We say that v covers u if u contains a subexpression $(xy)z$ (where x, y, z are bracketed strings) and v is obtained from u by replacing $(xy)z$ with $x(yz)$. For instance, $((a^2 \cdot a)a^2)(a^2 \cdot a^2)$ is covered by $((a \cdot a^2)a^2)(a^2 \cdot a^2)$, $(a^2(a \cdot a^2))(a^2 \cdot a^2)$, $((a^2 \cdot a)a^2)(a(a \cdot a^2))$, and $(a^2 \cdot a)(a^2(a^2 \cdot a^2))$. Figures 4.6 and 4.7 show the Hasse diagrams of T_3 and T_4. (In Figure 4.7, we have encoded the binary bracketing by a string of four $+$'s and four $-$'s, where a $+$ stands for a left parenthesis and a $-$ for the letter a, with the last a omitted.) Let U_n be the poset of all integer vectors $(a_1, a_2, \ldots, a_n)$ such that $i \leq a_i \leq n$ and such that if $i \leq j \leq a_i$ then $a_j \leq a_i$, ordered coordinatewise. Show that T_n and U_n are isomorphic posets.

(b) [2] Deduce from (a) that T_n is a lattice (called the *Tamari lattice*).

A42. Let C be a convex n-gon. Let $\mathcal{S}$ be the set of all sets of diagonals of C that do not intersect in the interior of C. Partially order the element of $\mathcal{S}$ by inclusion, and add a $\hat{1}$. Call the resulting poset A_n.

(a) [3–] Show that A_n is a simplicial Eulerian lattice of rank $n - 2$, as defined in [64, pp. 248, 310].

(b) [3] Show in fact that A_n is the lattice of faces of an $(n - 3)$-dimensional convex polytope $\mathcal{Q}_n$.

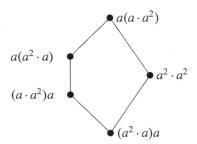

Figure 4.6. The Tamari lattice T_3.

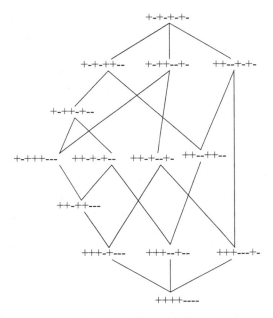

Figure 4.7. The Tamari lattice T_4.

(c) [3] For a complete binary tree T with $n + 1$ endpoints (item 5), let $V_T = (v_1, \ldots, v_n)$ be the internal vertices in left-to-right order, i.e., in the order in which they are visited in preorder by moving to that vertex in the direction of the root. Let a_i (respectively, b_i) be the number of leaves of the left subtree (respectively, right subtree) of v_i. Set $x_T = (a_1 b_1, \ldots, a_n b_n) \in \mathbb{R}^n$. For instance, the five trees of item 5 yield the points $(1, 2, 3)$, $(2, 1, 3)$, $(3, 1, 2)$, $(3, 2, 1)$, $(1, 4, 1)$. Show that the convex hull of the points x_T, where T ranges over all complete binary trees with $n + 1$ endpoints, is a convex polytope with face lattice isomorphic to A_n. Note that this construction gives an explicit answer to (b) above.

(d) [3–] Find the number $W_i = W_i(n)$ of elements of A_n of rank i. Equivalently, W_i is the number of ways to draw i diagonals of C that do not intersect in their interiors. Note that by [65, Prop. 6.2.1], $W_i(n)$ is also the number of plane trees with $n+i$ vertices and $n-1$ endpoints such that no vertex has exactly one child.

(e) [3–] Define

$$\sum_{i=0}^{n-3} W_i(x-1)^{n-i-3} = \sum_{i=0}^{n-3} h_i x^{n-3-i}, \qquad (4.13)$$

as in Equation (3.44). The vector $(h_0, \ldots, h_{n-3})$ is called the *h-vector* of A_n (or of the polytope $\mathcal{Q}_n$). Find an explicit formula for each h_i.

A43. There are many possible q-analogues of Catalan numbers. In (a) we give what is perhaps the most natural "combinatorial" q-analogue, while in (d) we give the most natural "explicit formula" q-analogue. In (e) we give an interesting extension of (d), while (f) and (g) are concerned with another special case of (e).

(a) [2+] Let

$$C_n(q) = \sum_P q^{A(P)},$$

where the sum is over all lattice paths P from $(0,0)$ to (n,n) with steps $(1,0)$ and $(0,1)$, such that P never rises above the line $y = x$, and where $A(P)$ is the area under the path (and above the x-axis). Note that by item 24, we have $C_n(1) = C_n$. (It is interesting to see what statistic corresponds to $A(P)$ for many of the other combinatorial interpretations of C_n given in Chapter 2.) For instance, $C_0(q) = C_1(q) = 1$, $C_2(q) = 1+q$, $C_3(q) = 1+q+2q^2+q^3$, $C_4(q) = 1+q+2q^2+3q^3+3q^4+3q^5+q^6$. Show that

$$C_{n+1}(q) = \sum_{i=0}^{n} C_i(q) C_{n-i}(q) q^{(i+1)(n-i)}.$$

Deduce that if $\tilde{C}_n(q) = q^{\binom{n}{2}} C_n(1/q)$, then the generating function

$$F(x) = \sum_{n \geq 0} \tilde{C}_n(q) x^n$$

satisfies

$$xF(x)F(qx) - F(x) + 1 = 0.$$

From this we get the continued fraction expansion

$$F(x) = \cfrac{1}{1 - \cfrac{x}{1 - \cfrac{qx}{1 - \cfrac{q^2 x}{1 - \cdots}}}}. \tag{4.14}$$

(b) [3] Define matrices $A_n = [\tilde{C}_{n+2-i-j}(q)]_{i,j=1}^{\lceil(n+1)/2\rceil}$. Show that A_n has a Smith normal form over $\mathbb{Z}[q]$ given by the diagonal matrix

$$\mathrm{diag}\left(q^{\binom{n}{2}}, q^{\binom{n-2}{2}}, q^{\binom{n-4}{2}}, \ldots, q^{\epsilon_n}\right),$$

where $\epsilon_n = q^{\binom{0}{2}} = 1$ if n is even, and $\epsilon_n = q^{\binom{1}{2}} = 1$ if n is odd. In particular,

$$\det A_n = \begin{cases} q^{\frac{1}{24}n(n+2)(2n-1)}, & n \text{ even} \\ q^{\frac{1}{24}(n-1)(n+1)(2n+3)}, & n \text{ odd}. \end{cases}$$

(c) [3] Let $A_n(q)$ be the number of right ideals of codimension n in the noncommutative polynomial ring $\mathbb{F}_q\langle x, y \rangle$. Show that $A_n(q) = q^{n(n+1)} C_n(1/q)$.

(d) [2+] Define

$$c_n(q) = \frac{1}{(n+1)} \binom{2n}{n},$$

where $\binom{2n}{n}$ denotes a q-binomial coefficient [64, §1.7]. For instance, $c_0(q) = c_1(q) = 1$, $c_2(q) = 1 + q^2$, $c_3(q) = 1 + q^2 + q^3 + q^4 + q^6$, $c_4(q) = 1 + q^2 + q^3 + 2q^4 + q^5 + 2q^6 + q^7 + 2q^8 + q^9 + q^{10} + q^{12}$. Show that

$$c_n(q) = \sum_w q^{\mathrm{maj}(w)},$$

where w ranges over all ballot sequences $a_1 a_2 \cdots a_{2n}$, and where

$$\mathrm{maj}(w) = \sum_{\{i : a_i > a_{i+1}\}} i,$$

the major index of w.

(e) [3–] Let t be a parameter, and define

$$c_n(t;q) = \frac{1}{(n+1)} \sum_{i=0}^{n} \binom{n}{i}\binom{n}{i+1} q^{i^2+it}.$$

Show that

$$c_n(t;q) = \sum_w q^{\mathrm{maj}(w)+(t-1)\mathrm{des}(w)},$$

where w ranges over the same set as in (b), and where

$$\mathrm{des}(w) = \#\{i : a_i > a_{i+1}\},$$

the number of descents of w. (Hence $c_n(1;q) = c_n(q)$.)

(f) [3–] Show that

$$c_n(0;q) = \frac{1+q}{1+q^n} c_n(q).$$

For instance, $c_0(0;q) = c_1(0;q) = 1$, $c_2(0;q) = 1+q$, $c_3(0;q) = 1+q+q^2+q^3+q^4$, $c_4(0;q) = 1+q+q^2+2q^3+2q^4+2q^5+2q^6+q^7+q^8+q^9$.

(g) [3+] Show that the coefficients of $c_n(0;q)$ are *unimodal*, i.e., if $c_n(0;q) = \sum b_i q^i$, then for some j we have $b_0 \le b_1 \le \cdots \le b_j \ge b_{j+1} \ge b_{j+2} \ge \cdots$. (In fact, we can take $j = \lfloor\frac{1}{2}\deg c_n(0;q)\rfloor = \lfloor\frac{1}{2}(n-1)^2\rfloor$.)

A44. (a) [4–] Let π be a "rotated Dyck path" of item 24. Define the *bounce path* of L to be the path described by the following algorithm. Start at $(0,0)$ and travel east along π until you encounter the beginning of a north step $(0,1)$. Then turn north and travel straight until you hit the diagonal $y = x$. Then turn east and travel straight until you again encounter the beginning of a north step of π, then turn north and travel to the diagonal, etc. Continue in this way until you arrive at (n,n). The bouncing ball will strike the diagonal at places $(0,0)$, (j_1,j_1), $(j_2,j_2),\ldots,(j_{b-1},j_{b-1})$, $(j_b,j_b) = (n,n)$. We define the *bounce* of π by

$$\mathrm{bounce}(\pi) = \sum_{i=1}^{b-1}(n - j_i).$$

Also define area(π) to be the area between the path L and the x-axis. Let

$$F_n(q,t) = \sum_\pi q^{\mathrm{area}(\pi)} t^{\mathrm{bounce}(\pi)},$$

where the sum is over all C_n lattice paths L of item 24. Let $B = A/I$ be the algebra of Problem A8(i). Then B has a natural $\mathbb{N}^2$-grading

$\bigoplus_{(i,j)\in\mathbb{N}^2} B_{ij}$ according to the x and y degrees. Let

$$G_n(q,t) = \sum_{i,j} \dim_{\mathbb{C}} B_{ij} q^i t^j,$$

the bigraded Hilbert series of B. Show that $F_n(q,t) = G_n(q,t)$.

(b) [5] Show combinatorially that $F_n(q,t) = F_n(t,q)$.

A45. Let Q_n be the poset of direct sum decompositions of an n-dimensional vector space V_n over the field $\mathbb{F}_q$, as defined in [65, Example 5.5.2(b)]. Let $\bar{Q}_n$ denote Q_n with a $\hat{0}$ adjoined, and let $\mu_n(q) = \mu_{\bar{Q}_n}(\hat{0},\hat{1})$, where $\mu_{\bar{Q}_n}$ denotes the Möbius function of $\bar{Q}_n$. Write

$$(n)! = 1 \cdot (1+q) \cdot (1+q+q^2) \cdots (1+q+\cdots+q^{n-1}). \tag{4.15}$$

Then by [65, (5.74)] we have

$$-\sum_{n\geq 1} \mu_n(q) \frac{x^n}{q^{\binom{n}{2}}(n)!} = \log \sum_{n\geq 0} \frac{x^n}{q^{\binom{n}{2}}(n)!}.$$

(a) [3−] Show that

$$\mu_n(q) = \frac{1}{n}(-1)^n (q-1)(q^2-1)\cdots(q^{n-1}-1)P_n(q),$$

where $P_n(q)$ is a polynomial in q of degree $\binom{n}{2}$ with nonnegative integral coefficients, satisfying $P_n(1) = \binom{2n-1}{n}$. For instance,

$$P_1(q) = 1$$
$$P_2(q) = 2+q$$
$$P_3(q) = 3+3q+3q^2+q^3$$
$$P_4(q) = (2+2q^2+q^3)(2+2q+2q^2+q^3).$$

(b) Show that

$$\exp \sum_{n\geq 1} q^{\binom{n}{2}} P_n(1/q) \frac{x^n}{n} = \sum_{n\geq 1} q^{\binom{n}{2}} C_n(1/q) x^n,$$

where $C_n(q)$ is the q-Catalan polynomial defined in Problem A43(a).

A46. (a) [2+] The *Narayana numbers* $N(n,k)$ are defined by

$$N(n,k) = \frac{1}{n}\binom{n}{k}\binom{n}{k-1}.$$

Let X_{nk} be the set of all ballot sequences $w = w_1 w_2 \cdots w_{2n}$ such that

$$k = \#\{j : w_j = 1, w_{j+1} = -1\}.$$

Give a combinatorial proof that $N(n,k) = \#X_{nk}$. Hence by item 77 there follows

$$\sum_{k=1}^{n} N(n,k) = C_n.$$

(It is interesting to find for each of the combinatorial interpretations of C_n given in Chapter 2 a corresponding decomposition into subsets counted by Narayana numbers.)

(b) [2+] Let $F(x,t) = \sum_{n\geq 1} \sum_{k\geq 1} N(n,k)x^n t^k$. Using the combinatorial interpretation of $N(n,k)$ given in (a), show that

$$xF^2 + (xt + x - 1)F + xt = 0, \tag{4.16}$$

so

$$F(x,t) = \frac{1 - x - xt - \sqrt{(1-x-xt)^2 - 4x^2 t}}{2x}.$$

A47. [3–] Write $\langle a \rangle_n = a(a+1)\cdots(a+n-1)$. Define

$$F_{a,b}(x) = \sum_{n\geq 0} \langle a \rangle_n \langle b \rangle_n \frac{x^n}{n!},$$

and set

$$\log F_{a,b}(x) = \sum_{n\geq 1} p_n(a,b) \frac{x^n}{n}.$$

Show that (a) $p_n(a,b)$ is a polynomial in a,b with nonnegative integer coefficients, (b) $p_n(a,b)$ has total degree $n+1$, and (c) the coefficient of $a^i b^{n+1-i}$ is the Narayana number $N(n,i)$.

A48. [2+] The *Motzkin numbers* M_n are defined by

$$\sum_{n\geq 0} M_n x^n = \frac{1 - x - \sqrt{1 - 2x - 3x^2}}{2x^2}$$

$$= 1 + x + 2x^2 + 4x^3 + 9x^4 + 21x^5 + 51x^6 + 127x^7 + 323x^8$$
$$+ 835x^9 + 2188x^{10} + \cdots.$$

Show that $M_n = \Delta^n C_1$ and $C_n = \Delta^{2n} M_0$, where C_n denotes a Catalan number and Δ is the difference operator. The difference tables that follow illustrate this result.

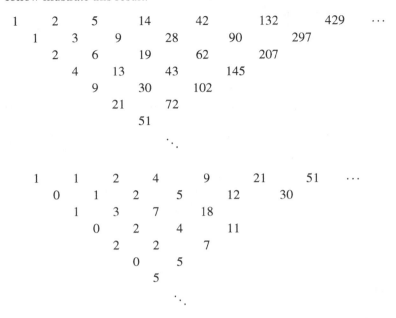

```
1    2    5    14      42        132        429    ...
   1    3    9      28        90         297
      2    6    19      62        207
         4    13      43        145
            9    30      102
              21      72
                51
                   ⋱
```

```
1    1    2    4      9      21      51    ...
   0    1    2      5      12      30
      1    3    7      18
         0    2    4      11
            2    2      7
              0      5
                5
                   ⋱
```

A49. [3–] Show that the Motzkin number M_n has the following interpretations as the number of elements of the given set. (See Problem A54(b) for an additional interpretation.)

(a) Ways of drawing any number of nonintersecting chords among n points on a circle.

(b) Walks on $\mathbb{N}$ with n steps, with steps -1, 0, or 1, starting and ending at 0.

(c) Lattice paths from $(0,0)$ to (n,n), with steps $(0,2)$, $(2,0)$, and $(1,1)$, never rising above the line $y = x$.

(d) Paths from $(0,0)$ to $(n,0)$ with steps $(1,0)$, $(1,1)$, and $(1,-1)$, never going below the x-axis. Such paths are called *Motzkin paths*.

(e) Pairs $1 \le a_1 < \cdots < a_k \le n$ and $1 \le b_1 < \cdots < b_k \le n$ of integer sequences such that $a_i \le b_i$ and every integer in the set $[n]$ appears at least once among the a_i's and b_i's.

(f) Ballot sequences (as defined in Section 1.5) $(a_1, \ldots, a_{2n+2})$ such that we never have $(a_{i-1}, a_i, a_{i+1}) = (1, -1, 1)$.

(g) Plane trees with $n/2$ edges, allowing "half edges" that have no children and count as half an edge.

(h) Plane trees with $n + 1$ edges in which no vertex, the root excepted, has exactly one child.

(i) Plane trees with n edges in which every vertex has at most two children.

(j) Binary trees with $n - 1$ edges such that no two consecutive edges slant to the right.

(k) Plane trees with $n + 1$ vertices such that every vertex of odd height (with the root having height 0) has at most one child.

(l) Noncrossing partitions $\pi = \{B_1, \ldots, B_k\}$ of $[n]$ (as defined in item 159) such that if $B_i = \{b\}$ and $a < b < c$, then a and c appear in different blocks of π.

(m) Noncrossing partitions π of $[n + 1]$ such that no block of π contains two consecutive integers.

(n) 2143-avoiding involutions (or vexillary involutions) in $\mathfrak{S}_n$.

(o) Binary trees with $n + 1$ vertices such that every non-endpoint vertex has a nonempty right subtree.

(p) Permutations $w \in \mathfrak{S}_n$ such that both w and w^{-1} have genus 0, as defined in item 118.

A50. [3–] The Schröder numbers[3] r_n and s_n are discussed in [65, §6.2]. They can be defined as follows:

$$\sum_{n \geq 0} s_n x^n = \frac{1 + x - \sqrt{1 - 6x + x^2}}{4x}$$

$$= 1 + x + 3x^2 + 11x^3 + 45x^4 + 197x^5 + 903x^6 + 4279x^7$$
$$+ 20793x^8 + 103049x^9 + \cdots$$

$$r_n = 2s_n, \ n \geq 1.$$

Show that the Schröder number have the following combinatorial interpretations.

(a) s_{n-1} is the total number of bracketings (parenthesizations) of a string of n letters.

(b) s_{n-1} is the number of plane trees with no vertex of degree one and with n endpoints.

(c) r_{n-1} is the number of plane trees with n vertices and with a subset of the endpoints circled.

(d) s_n is the number of binary trees with n vertices and with each right edge colored either red or blue.

---

[3] It is interesting that the Schröder numbers were known to the Greek astronomer Hipparchus (c. 190–c. 120 BCE). See R. Stanley, *Amer. Math. Monthly* **104** (1997), 344–350, and F. Acerbi, *Archive for History of Exact Sciences* **57** (2003), 465–502.

(e) s_n is the number of lattice paths in the (x, y) plane from $(0, 0)$ to the x-axis using steps $(1, k)$, where $k \in \mathbb{P}$ or $k = -1$, never passing below the x-axis, and with n steps of the form $(1, -1)$.

(f) s_n is the number of lattice paths in the (x, y) plane from $(0, 0)$ to (n, n) using steps $(k, 0)$ or $(0, 1)$ with $k \in \mathbb{P}$, and never passing above the line $y = x$.

(g) r_n is the number of paths from $(0, 0)$ to $(2n, 0)$ with steps $(1, 1)$, $(1, -1)$, and $(2, 0)$, never passing below the x-axis.

(h) s_n is the number of paths in (g) with no level steps on the x-axis.

(i) r_{n-1} is the number of parallelogram polyominoes (defined in the solution to item 57) of perimeter $2n$ with each column colored either black or white.

(j) s_n is the number of ways to draw any number of diagonals of a convex $(n + 2)$-gon that do not intersect in their interiors.

(k) s_n is the number of sequences $i_1 i_2 \cdots i_k$, where $i_j \in \mathbb{P}$ or $i_j = -1$ (and k can be arbitrary), such that $n = \#\{j : i_j = -1\}$, $i_1 + i_2 + \cdots + i_j \geq 0$ for all j, and $i_1 + i_2 + \cdots + i_k = 0$.

(l) r_n is the number of lattice paths from $(0, 0)$ to (n, n), with steps $(1, 0)$, $(0, 1)$, and $(1, 1)$, that never rise above the line $y = x$.

(m) r_{n-1} is the number of $n \times n$ permutation matrices P with the following property: we can eventually reach the all 1's matrix by starting with P and continually replacing a 0 by a 1 if that 0 has at least two adjacent 1's, where an entry a_{ij} is defined to be adjacent to $a_{i \pm 1, j}$ and $a_{i, j \pm 1}$.

(n) Let $\mathfrak{S}_n(u, v)$ denote the set of permutations $w \in \mathfrak{S}_n$ avoiding both the permutations $u, v \in \mathfrak{S}_4$. There is a group G of order 16 that acts on the set of pairs (u, v) of unequal elements of $\mathfrak{S}_4$ such that if (u, v) and (u', v') are in the same G-orbit (in which case we say that they are *equivalent*), then there is a simple bijection between $\mathfrak{S}_n(u, v)$ and $\mathfrak{S}_n(u', v')$ (for all n). Namely, identifying a permutation with the corresponding permutation matrix, the orbit of (u, v) is obtained by possibly interchanging u and v, and then doing a simultaneous dihedral symmetry of the square matrices u and v. There are then ten inequivalent pairs $(u, v) \in \mathfrak{S}_4 \times \mathfrak{S}_4$ for which $\#\mathfrak{S}_n(u, v) = r_{n-1}$, namely, $(1234, 1243)$, $(1243, 1324)$, $(1243, 1342)$, $(1243, 2143)$, $(1324, 1342)$, $(1342, 1423)$, $(1342, 1432)$, $(1342, 2341)$, $(1342, 3142)$, and $(2413, 3142)$.

(o) r_{n-1} is the number of permutations $w = w_1 w_2 \cdots w_n$ of $[n]$ with the following property: it is possible to insert the numbers $w_1, \ldots, w_n$ in order into a string, and to remove the numbers from the string in the order $1, 2, \ldots, n$. Each insertion must be at the beginning or end of the string. At any time we may remove the first (leftmost) element of the string, so long as we remove the numbers in the order $1, 2, \ldots, n$. (*Example:* $w = 2413$. Insert 2, insert 4 at the right, insert 1 at the left, remove 1, remove 2, insert 3 at the left, remove 3, remove 4.)

(p) s_n is the number of pairs $((a_1, a_2, \ldots, a_k), (i_1, i_2, \ldots, i_k)) \in \mathbb{P}^k \times \mathbb{P}^k$ satisfying $1 = a_1 < a_2 < \cdots < a_k \le n$, $1 \le i_1 \le i_2 \le \cdots \le i_k$, and $i_j \le a_j$.

(q) r_n is the number of sequences of length $2n$ from the alphabet A, B, C such that: (i) for every $1 \le i < 2n$, the number of A's and B's among the first i terms is not less than the number of C's, (ii) the total number of A's and B's is n (and hence also the total number of C's), and (iii) no two consecutive terms are of the form CB.

(r) r_{n-1} is the number of noncrossing partitions (as defined in item 159) of some set $[k]$ into n blocks, such that no block contains two consecutive integers.

(s) s_n is the number of graphs G (without loops and multiple edges) on the vertex set $[n+2]$ with the following two properties: (α) All of the edges $\{1, n+2\}$ and $\{i, i+1\}$ are edges of G, and (β) G is *noncrossing*, i.e., there are not both edges $\{a, c\}$ and $\{b, d\}$ with $a < b < c < d$. Note that an arbitrary noncrossing graph on $[n+2]$ can be obtained from those satisfying (α)–(β) by deleting any subset of the required edges in (α). Hence the total number of noncrossing graphs on $[n+2]$ is $2^{n+2} s_n$.

(t) r_{n-1} is the number of reflexive and symmetric relations R on the set $[n]$ such that if iRj with $i < j$, then we never have uRv for $i \le u < j < v$.

(u) r_{n-1} is the number of reflexive and symmetric relations R on the set $[n]$ such that if iRj with $i < j$, then we never have uRv for $i < u \le j < v$.

(v) r_{n-1} is the number of ways to cover with disjoint dominos (or dimers) the set of squares consisting of $2i$ squares in the ith row for $1 \le i \le n-1$, and with $2(n-1)$ squares in the nth row, such that the row centers lie on a vertical line. See Figure 4.8 for the case $n = 4$.

(w) s_n is the number of $2 \times n$ matrices whose set of entries (ignoring multiplicities) is $1, 2, \ldots, k$ for some k, such that every row and column is strictly increasing.

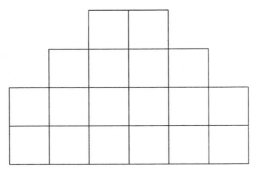

Figure 4.8. A board with $r_3 = 22$ domino tilings.

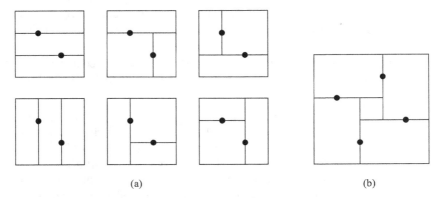

(a)   (b)

Figure 4.9. Dividing a rectangle into rectangles.

(x) Given n points $p_1,\dots,p_n$ in the interior of a rectangle R with sides
    parallel to the x and y-axes, such that no two of the points are
    parallel to the x-axis or y-axis, r_n is the number of *rectangulations*
    (ways to divide R into rectangles) so that every p_i lies on an
    edge of some rectangle, and so that the process of dividing R
    can be carried out in n steps, each step dividing some rectangle
    into two. Such rectangulations are called *guillotine* rectangulations.
    The total number of rectangles into which R is divided will be
    $n + 1$. Figure 4.9(a) shows the $r_2 = 6$ ways when $n = 2$, while
    Figure 4.9(b) shows a non-guillotine rectangulation.

(y) Conjecturally, r_{n-1} is the number of permutations $w \in \mathfrak{S}_n$ (under
    the weak order) for which

$$F(\Lambda_w, q)F(V_w, q) = (n)!,$$

where $\Lambda_w = \{u : u \leq w\}$, $V_w = \{v : v \geq w\}$, and $F(P,q)$ denotes the rank-generating function of the poset P. The symbol $(n)!$ is defined in Equation (4.15).

(z) r_n is the number of $(n+1) \times (n+1)$ alternating sign matrices A (defined in item 209) such that we never have $A_{ii'} = A_{jj'} = A_{kk'} = 1$, where $i < j < k$ and $i' < k' < j'$ (such matrices are called 132-*avoiding* alternating sign matrices).

(aa) r_{n-1} is the number of permutations $w \in \mathfrak{S}_n$ such that there exist real polynomials $p_1(x),\ldots,p_n(x)$, all vanishing at $x = 0$, with the following property: if $p_1(x) < p_2(x) < \cdots < p_n(x)$ for small (i.e., sufficiently close to 0) $x < 0$, then $p_{w(1)}(x) < p_{w(2)}(x) < \cdots < p_{w(n)}(x)$ for small $x > 0$.

A51. [3–] Let a_n be the number of permutations $w = w_1 w_2 \cdots w_n \in \mathfrak{S}_n$ such that we never have $w_{i+1} = w_i \pm 1$, e.g., $a_4 = 2$, corresponding to 2413 and 3142. Equivalently, a_n is the number of ways to place n nonattacking kings on an $n \times n$ chessboard with one king in every row and column. Let

$$A(x) = \sum_{n \geq 0} a_n x^n$$
$$= 1 + x + 2x^4 + 14x^5 + 90x^6 + 646x^7 + 5242x^8 + \cdots.$$

Show that $A(xR(x)) = \sum_{n \geq 0} n! x^n := E(x)$, where

$$R(x) = \sum_{n \geq 0} r_n x^n = \frac{1}{2x}\left(1 - x - \sqrt{1 - 6x + x^2}\right),$$

the generating function for Schröder numbers. Deduce that

$$A(x) = E\left(\frac{x(1-x)}{1+x}\right).$$

A52. [3] A permutation $w \in \mathfrak{S}_n$ is called *2-stack sortable* if $S^2(w) = w$, where S is the operator of item 119. Show that the number $S_2(n)$ of 2-stack sortable permutations in $\mathfrak{S}_n$ is given by

$$S_2(n) = \frac{2(3n)!}{(n+1)!(2n+1)!}.$$

A53. Define a *Catalan triangulation* of the Möbius band to be an abstract simplicial complex triangulating the Möbius band that uses no interior vertices, and has vertices labeled $1, 2, \ldots, n$ in order as one traverses the boundary. (If we replace the Möbius band by a disk, then we get the triangulations of Theorem 1.5.1(i) or item 1.) Figure 4.10 shows

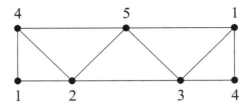

Figure 4.10. A Catalan triangulation of the Möbius band.

the smallest such triangulation, with five vertices (where we identify the vertical edges of the rectangle in opposite directions). Let $MB(n)$ be the number of Catalan triangulations of the Möbius band with n vertices. Show that

$$\sum_{n\geq 0} MB(n)x^n = \frac{x^2\left((2-5x-4x^2)+(-2+x+2x^2)\sqrt{1-4x}\right)}{(1-4x)\left(1-4x+2x^2+(1-2x)\sqrt{1-4x}\right)}$$

$$= x^5 + 14x^6 + 113x^7 + 720x^8 + 4033x^9 + 20864x^{10} + \cdots.$$

A54. (a) [3+] Let $f(n)$ denote the number of subsets S of $\mathbb{N} \times \mathbb{N}$ of cardinality n with the following property: if $p \in S$, then there is a lattice path from $(0,0)$ to p with steps $(0,1)$ and $(1,0)$, all of whose vertices lie in S. Show that

$$\sum_{n\geq 1} f(n)x^n = \frac{1}{2}\left(\sqrt{\frac{1+x}{1-3x}}-1\right)$$

$$= x + 2x^2 + 5x^3 + 13x^4 + 35x^5 + 96x^6 + 267x^7$$
$$+ 750x^8 + 2123x^9 + 6046x^{10} + \cdots.$$

    (b) [3+] Show that the number of such subsets contained in the first octant $0 \leq x \leq y$ is the Motzkin number M_{n-1} (defined in Problem A48).

A55. (a) [3] Let P_n be the Bruhat order on the symmetric group $\mathfrak{S}_n$ as defined in [64, Exercise 3.183]. Show that the following two conditions on a permutation $w \in \mathfrak{S}_n$ are equivalent:

    i. The interval $[\hat{0}, w]$ of P_n is rank-symmetric, i.e., if ρ is the rank function of P_n (so $\rho(w)$ is the number of inversions of w), then

$$\#\{u \in [\hat{0}, w] : \rho(u) = i\} = \#\{u \in [\hat{0}, w] : \rho(w) - \rho(u) = i\},$$

    for all $0 \leq i \leq \rho(w)$.

    ii. The permutation $w = w_1 w_2 \cdots w_n$ is 4231 and 3412-avoiding, i.e., there do not exist $a < b < c < d$ such that $w_d < w_b < w_c < w_a$ or $w_c < w_d < w_a < w_b$.

(b) [3–] Call a permutation $w \in \mathfrak{S}_n$ *smooth* if it satisfies (i) (or (ii)) above. Let $f(n)$ be the number of smooth $w \in \mathfrak{S}_n$. Show that

$$\sum_{n \geq 0} f(n) x^n = \frac{1}{1 - x - \dfrac{x^2}{1-x} \left(\dfrac{2x}{1+x-(1-x)C(x)} - 1 \right)}$$

$$= 1 + x + 2x^2 + 6x^3 + 22x^4 + 88x^5 + 366x^6$$
$$+ 1552x^7 + 6652x^8 + 28696x^9 + \cdots .$$

A56. [2] Let b_n be the number of ways of parenthesizing a string of n letters, subject to a *commutative* (but nonassociative) binary operation. Thus for instance $b_5 = 3$, corresponding to the parenthesizations

$$x^2 \cdot x^3 \qquad x \cdot (x \cdot x^3) \qquad x(x^2 \cdot x^2).$$

(Note that x^3 is unambiguous, since $x \cdot x^2 = x^2 \cdot x$.) Let

$$B(x) = \sum_{n \geq 1} b_n x^n$$

$$= x + x^2 + x^3 + 2x^4 + 3x^5 + 6x^6 + 11x^7 + 23x^8 + 46x^9 + 98x^{10} + \cdots .$$

Show that $B(x)$ satisfies the functional equation

$$B(x) = x + \frac{1}{2}B(x)^2 + \frac{1}{2}B(x^2). \tag{4.17}$$

A57. [3] Let G be a finite group with identity e and integral group ring $\mathbb{Z}G$. Let T be a subset of G, and write $\widetilde{T} = \sum_{g \in T} g \in \mathbb{Z}G$. The elements $\widetilde{T}$ are called *simple quantities*. Define $\widetilde{T}^t = \sum_{g \in T} g^{-1}$. A *Schur ring* over G is a subring $\mathcal{S}$ of $\mathbb{Z}G$ that is generated as a $\mathbb{Z}$-module by simple quantities $\widetilde{T}_1, \ldots, \widetilde{T}_r$, called the *basic sets* of $\mathcal{S}$, satisfying the axioms:

- $T_1 = \{e\}$
- $T_1, \ldots, T_r$ form a partition of G,
- For each i there is a j such that $\widetilde{T}_i^t = \widetilde{T}_j$.

Let $\mathcal{S}$ be a Schur ring over $G = \mathbb{Z}/n\mathbb{Z}$. We say that $\mathcal{S}$ is *decomposable* if there is a nontrivial, proper subgroup H of $\mathbb{Z}/n\mathbb{Z}$ such that for every basic set T, either $T \subseteq H$ or $T = \bigcup_{h \in T} Hh$; otherwise $\mathcal{S}$ is *indecomposable*. Show that if p is an odd prime, then the number of indecomposable Schur rings over $\mathbb{Z}/p^m\mathbb{Z}$ is equal to $\tau(p-1)C_{m-1}$, where $\tau(p-1)$ is the number of (positive) divisors of $p-1$.

A58. This problem assumes a knowledge of matroid theory. Given a Dyck path $P = a_1 a_2 \cdots a_{2n}$ of length $2n$ (where each $a_i = U$ or D), let $\mathcal{U}(P) = \{i : a_i = U\}$, the *up-step set* of P.

(a) [2+] Show that the set $\mathcal{B}_n$ of the up-step sets of all Dyck paths of length $2n$ is the collection of bases of a matroid C_n on the set $[2n]$, called the *Catalan matroid*.

(b) [2] Show that C_n is self-dual.

(c) [2+] Let $a(P)$ denote the number of up steps of P before the first down step, and let $b(P)$ denote the number of times the Dyck path returns to the x-axis after its first up step. Show that the Tutte polynomial of C_n is given by

$$T_{C_n}(q,t) = \sum_P q^{a(P)} t^{b(P)},$$

where the sum is over all Dyck paths P of length $2n$.

(d) [2+] Show that

$$\sum_{n \geq 0} T_{C_n}(q,t) x^n = \frac{1 + (qt - q - t)xC(x)}{1 - qtx + (qt - q - t)xC(x)}.$$

A59. (a) [3−] Find the unique continuous function $f(x)$ on $\mathbb{R}$ satisfying for all $n \in \mathbb{N}$:

$$\int_{-\infty}^{\infty} x^n f(x)\,dx = \begin{cases} C_k, & \text{if } n = 2k \\ 0, & \text{if } n = 2k+1. \end{cases}$$

(b) [3−] Find the unique continuous function $f(x)$ for $x > 0$ satisfying for all $n \in \mathbb{N}$:

$$\int_0^{\infty} x^n f(x)\,dx = C_n.$$

A60. [3] Show that

$$C_{n-1} = \sum \prod_{i=1}^{k} \left(\binom{a_{i-1} + a_i - 1}{a_i} - 1 \right),$$

where the sum is over all compositions $a_0 + a_1 + \cdots + a_k = n$. The composition with $k = 0$ contributes 1 to the sum.

A61. (a) [2] Let $b(n)$ denote the number of 1's in the binary expansion of n. Using Kummer's theorem on binomial coefficients modulo a prime power, show that the exponent of the largest power of 2 dividing C_n is equal to $b(n+1) - 1$.

(b) [3−] Give a combinatorial proof.

(c) [3] Let $\Phi(x) = \sum_{n \geq 0} x^{2^n}$. Show that

$$\sum_{n \geq 0} C_n x^n \equiv 32x^5 + 16x^4 + 6x^2 + 13x + 1$$

$$+ (32x^4 + 32x^3 + 20x^2 + 44x + 40)\Phi(x)$$

$$+ \left(16x^3 + 56x^2 + 30x + 52 + \frac{12}{x}\right)\Phi(x)^2$$

$$+ \left(32x^3 + 60x + 60 + \frac{28}{x}\right)\Phi(x)^3$$

$$+ \left(32x^3 + 16x^2 + 48x + 18 + \frac{35}{x}\right)\Phi(x)^4$$

$$+ (32x^2 + 44)\Phi(x)^5$$

$$+ \left(48x + 8 + \frac{50}{x}\right)\Phi(x)^6$$

$$+ \left(32x + 32 + \frac{4}{x}\right)\Phi(x)^7 \pmod{64},$$

where the congruence modulo 64 holds coefficientwise.

A62. Like the previous problem, for this problem it may also be useful to use Kummer's theorem.

(a) [2+] For every integer $k \neq 1$, show that there are infinitely many positive integers n for which $\binom{2n}{n}$ is not divisible by $n + k$. Thus the situation for $k = 1$, where $\frac{1}{n+1}\binom{2n}{n} = C_n$, is quite special.

(b) [3] A set S of positive integers has *asymptotic density* 1 if

$$\lim_{n \to \infty} \frac{\#(S \cap [n])}{n} = 1.$$

Show that for each positive integer k, the set of positive integers n for which $\binom{2n}{n}$ is divisible by $(n+1)(n+2) \cdots (n+k)$ has asymptotic density 1.

(c) [3] A set S of positive integers has *upper asymptotic density* α if

$$\limsup_{n \to \infty} \frac{\#(S \cap [n])}{n} = \alpha.$$

For each $k \geq 0$ show that the set of integers $n > k$ for which $\binom{2n}{n}$ is divisible by $n - k$ is infinite, but has upper asymptotic density smaller than $\frac{1}{3}$.

A63. [3–] Let $f(n)$ be the number of Catalan numbers C_i, $1 \leq i \leq n$, that are a sum of three squares. Show that

$$\lim_{n\to\infty} \frac{f(n)}{n} = \frac{7}{8}.$$

Compare with the known result that if $g(n)$ is the number of integers $1 \leq i \leq n$ that are a sum of three squares, then $\lim_{n\to\infty} \frac{g(n)}{n} = \frac{5}{6}$.

NOTE. A positive integer N is not a sum of three squares if and only if there exist $a, b \in \mathbb{N}$ for which $N = 4^a(8b+7)$.

A64. [2–] Write $f(n) \sim g(n)$ to denote that $\lim_{n\to\infty} \frac{f(n)}{g(n)} = 1$. Assuming Stirling's formula $n! \sim n^n e^{-n} \sqrt{2\pi n}$, show that

$$C_n \sim \frac{4^n}{\sqrt{\pi} n^{3/2}}.$$

A65. (a) [3–] Show that

$$\sum_{n\geq 0} \frac{x^n}{C_n} = \frac{2(x+8)}{(4-x)^2} + \frac{24\sqrt{x}\sin^{-1}\left(\frac{1}{2}\sqrt{x}\right)}{(4-x)^{5/2}}. \qquad (4.18)$$

(b) [1+] Deduce that

$$\sum_{n\geq 0} \frac{1}{C_n} = 2 + \frac{4\sqrt{3}\pi}{27}.$$

(c) [2+] Show that

$$\sum_{n\geq 0} \frac{4-3n}{C_n} = 2.$$

A66. [2] Show that $\sum_{n\geq 0} \dfrac{C_n}{4^n} = 2$.

HINT. Use Abel's theorem, which asserts that if $F(x) = \sum_{n\geq 0} a_n x^n \in \mathbb{C}[[x]]$ and $\sum_{n\geq 0} a_n$ converges to L, then

$$\lim_{x\to 1^-} F(x) = L,$$

where x assumes only real values as it approaches 1 from below.

A67. (a) [2+] Show that

$$\log C(x) = \sum_{n\geq 1} \binom{2n-1}{n} \frac{x^n}{n}. \qquad (4.19)$$

(b) [3–] Give a combinatorial proof based on item 6 and the exponential formula [65, Cor. 5.1.6].

A68. [3–] Euler found in 1737 the continued fraction expansion

$$e = 1 + \cfrac{2}{1 + \cfrac{1}{6 + \cfrac{1}{10 + \cfrac{1}{14 + \cfrac{1}{18 + \cfrac{1}{22 + \cdots}}}}}}.$$

The convergents to this continued fraction are $1, 3, \frac{19}{7}, \frac{193}{71}, \ldots$. Let a_n be the numerator of the nth convergent, so $a_1 = 1$, $a_2 = 3$, $a_3 = 19$, $a_4 = 193$. Show that

$$1 + \sum_{n \geq 1} a_n \frac{x^n}{n!} = \exp xC(x).$$

5

Solutions to Additional Problems

A1. For a generating function proof, set $y = \sum_{n \geq 1} D_n x^n$. Then $y = y^2 + x$, giving $y = \frac{1}{2}(1 - \sqrt{1 - 4x}) = xC(x)$, so $D_n = C_{n-1}$. There are numerous other ways to see this result.

A2. (a) Given a path P of the first type, let (i, i) be the first point on P that intersects $y = x$. Replace the portion of P from $(1, 0)$ to (i, i) by its reflection about $y = x$. This yields the desired bijection.

This argument is the famous *reflection principle*, often attributed to D. André. Although André did provide a combinatorial proof of a generalization of the enumeration of ballot sequences [3], he did not actually use the reflection principle. For a discussion of the history of the reflection principle, see Section B.7. The importance of the reflection principle in combinatorics and probability theory was realized by W. Feller, *An Introduction to Probability Theory and Its Applications*, vol. 1, John Wiley and Sons, New York, 1950 (3rd edition, 1968). For a vast number of extensions and ramifications, see L. Takács, *Combinatorial Methods in the Theory of Stochastic Processes*, John Wiley and Sons, New York, 1967; T. V. Narayana [52]; and S. G. Mohanty [51]. For a profound generalization of the reflection principle based on the theory of Coxeter groups, as well as some additional references, see I. Gessel and D. Zeilberger, *Proc. Amer. Math. Soc.* **115** (1992), 27–31.

(b) The first step in such a lattice path must be from $(0, 0)$ to $(1, 0)$. Hence we must subtract from the total number of paths from $(1, 0)$ to (m, n) the number that intersect $y = x$, so by (a) we get $\binom{m+n-1}{n} - \binom{m+n-1}{m} = \frac{m-n}{m+n}\binom{m+n}{n}$.

(c) Move the path one unit to the right to obtain the case $m = n + 1$ of (b).

(d) Straightforward from (b). The result is known as *Bertrand's ballot theorem* and was first published by W. A. Whitworth in 1878. Bertrand rediscovered it in 1887. Pak has observed that the numbers $B(m,n)$ appear in the paper by E. Catalan, *J. Math. Pures et Appliquées* **4** (1839), 91–94, where they are denoted $P_{m+1,m-3-n}$ (with $B(m,m-1)$ and $B(m,m-2)$ denoted P_{m+2}). Catalan interprets $P_{m,i}$ as the number of triangulations T of an $(n+1)$-gon with vertices $1,2,\ldots,n+1$ in that order such that i is the least integer for which $i+2,i+3,i+4$ are the vertices of a triangle in T. For further information, see Section B.7.

(e) Easy to deduce from item 170, for instance.

A3. (a) Given a path $P \in X_n$, define $c(P) = (c_0,c_1,\ldots,c_n)$, where c_i is the number of horizontal steps of P at height $y = i$. It is not difficult to verify that the cyclic permutations $K_j = (c_j,c_{j+1},\ldots,c_n,c_1,\ldots,c_{j-1})$ of $c(P)$ are all distinct, and for each such there is a unique $P_j \in X_n$ with $c(P_j) = K_j$. Moreover, the number of excedances of the paths $P = P_0, P_1, \ldots, P_n$ are just the numbers $0, 1, \ldots, n$ in some order. From these observations the proof is immediate.

This result, known as the *Chung-Feller theorem*, is due to K. L. Chung and W. Feller, *Proc. Nat. Acad. Sci. U.S.A.* **35** (1949), 605–608. A refinement was given by T. V. Narayana, *Skand. Aktuarietidskr.* (1967), 23–30. For further information see Narayana [52, §I.2] and Mohanty [51, §3.3]. For another generalization, see A. Huq, PhD thesis, Brandeis University, 2009; arXiv:0907.3254.

(b) Immediate from (a).

A4. This result was given by L. W. Shapiro, problem E2903, *Amer. Math. Monthly* **88** (1981), 619. An incorrect solution appeared in **90** (1983), 483–484. A correct but nonbijective solution was given by D. M. Bloom, **92** (1985), 430. The editors asked for a bijective proof in problem E3096, **92** (1985), 428, and such a proof was given by W. Nichols, **94** (1987), 465–466. Nichols's bijection is the following. Regard a lattice path as a sequence of E's (for the step $(1,0)$) and N's (for the step $(0,1)$). Given a path P of the type we are enumerating, define recursively a new path $\psi(P)$ as follows:

$$\psi(\emptyset) = \emptyset, \quad \psi(DX) = D\psi(X), \quad \psi(D'X) = E\psi(X)ND^*,$$

where (a) D is a path of positive length, with endpoints on the diagonal $x = y$ and all other points below the diagonal, (b) D' denotes the path obtained from D by interchanging E's and N's, and (c) $D = ED^*N$. Then ψ establishes a bijection between the paths we are enumerating and the paths of item 24 with n replaced by $2n$. For an explicit description of ψ^{-1} and a proof that ψ is indeed a bijection, see the solution of Nichols cited above.

A5. (a) *First solution* (sketch). We count the number $s(n)$ of triangles that are *not* internal. Each edge of $\mathcal{P}_{n+2}$ occur in $\sum C_i C_{n-1-i} = C_n$ non-internal triangles. Each internal triangle has exactly one edge on $\mathcal{P}_{n+2}$, except that $n+2$ of them have two edges in $\mathcal{P}_{n+2}$. There are C_{n-1} triangulations which contain a fixed triangle with two edges on $\mathcal{P}_{n+2}$. Hence

$$t(n) = nC_n - ((n+2)C_n - (n+2)C_{n-1}) = (n+2)C_{n-1} - 2C_n.$$

Second solution (sketch). In the bijection between triangulations and binary trees in the proof of Theorem 1.5.1, internal triangles correspond to vertices of total degree three (that is, with a parent and two children). Let

$$y_i = \sum_T t^{d(T)} x^{v(T)},$$

where T ranges over all binary trees for which the root has i children, also allowing the one-vertex tree when $i = 1$, and where $d(T)$ is the number of vertices of T of total degree three, and $v(T)$ is the number of vertices of T. Then

$$y_1 = x + 2xy_1 + 2xty_2$$
$$y_2 = x(y_1 + yy_2)^2.$$

Solving gives that $y_1 + y_2$ is equal to

$$\frac{1 - 2x(2-t) + 2x^2(1-t)(2-t) - (1 - 2x(1-t))\sqrt{1 - 4x + 4(1-t)x^2}}{2t^2 x}.$$

Now compute $\left.\frac{\partial(y_1+y_2)}{\partial t}\right|_{t=1}$ and simplify.

A6. The Black pawn on a5 must promote to a bishop and move to a7 without checking White or capturing the pawn at b6. This takes 17 moves, which are unique. The Black pawn on a6 must follow in the footsteps of the first Black pawn and end up at b8. This also takes 17 moves, which are unique. (A knight promotion won't do because a knight at a1 cannot move without checking White.) White then plays Pb7 mate. Denote a move by the pawn on a5 by $+1$ and a move by the pawn on a6 by -1. Since the pawn on a6 can never overtake the pawn on a5 (even after promotion), it follows that the number of solutions is just the number of sequences of seventeen 1's and seventeen -1's with all partial sums nonnegative. By item 77, the number of solutions is therefore the Catalan number $C_{17} = 129644790$.

Why the configuration of five Black men in the lower right-hand corner? A Black pawn is needed at h4 to prevent a bishop path going through h4. It wouldn't do simply to place a Black pawn at h4 and White pawn at h3, since then there will be shorter solutions obtained by Black capturing the pawn at h3 with a queen and promoting the pawn at h4 to a knight. Note that the five men in the lower right-hand corner are permanently locked in place, since any move by the Black knight would check White. We leave the reader to ponder the reason for the White pawn at c5.

This problem is due to R. Stanley, *Suomen Tehtäväniekat* **59**, no. 4 (2005), 193–203 (Problem (B)), based on an earlier problem of K. Väisänen (with only C_9 solutions) which appears in A. Puusa, *Queue Problems*, Finnish Chess Problem Society, Helsinki, 1992 (Problem 2). This booklet contains fifteen problems of a similar nature. See also [65, Exercise 7.18]. For more information on serieshelpmates in general, see A. Dickins, *A Guide to Fairy Chess*, Dover, New York, 1971, p. 10; and J. M. Rice and A. Dickins, *The Serieshelpmate*, 2nd edition, Q Press, Kew Gardens, 1978.

A7. These are just Catalan numbers! See for instance J. Gili, *Catalan Grammar*, Dolphin, Oxford, 1993, p. 39. A related question appears in *Amer. Math. Monthly* **103** (1996), 538 and 577.

A8. (a) Follows from [64, Exercise 3.76(b)] and item 178. See L. W. Shapiro, *American Math. Monthly* **82** (1975), 634–637.

(b) We assume knowledge of Chapter 7 of [65]. It follows from the results of Appendix 2 of Chapter 7 that we want the coefficient of the trivial Schur function $s_\emptyset$ in the Schur function expansion of $(x_1 + x_2)^{2n}$ in the ring $\Xi_2 = \Lambda_2/(x_1 x_2 - 1)$. Since $s_\emptyset = s_{(n,n)}$ in Ξ_2, the number we want is just $\langle s_1^{2n}, s_{(n,n)} \rangle = f^{(n,n)}$ (using [65, Cor. 7.103]), and the result follows from item 168.

(c) See R. Stanley, *Ann. New York Acad. Sci.*, vol. 576, 1989, pp. 500–535 (Example 4 on p. 523).

(d) See R. Stanley, in *Advanced Studies in Pure Math.*, vol. 11, Kinokuniya, Tokyo, and North-Holland, Amsterdam/New York, 1987, pp. 187–213 (bottom of p. 194). A simpler proof follows from R. Stanley, *J. Amer. Math. Soc.* **5** (1992), 805–851 (Prop. 8.6). For a related result, see C. Chan, *SIAM J. Disc. Math.* **4** (1991), 568–574.

(e) See L. R. Goldberg, *Adv. Math.* **85** (1991), 129–144 (Thm. 1.7).

(f) See D. Tischler, *J. Complexity* **5** (1989), 438–456.

(g) This algebra is the *Temperley-Lieb algebra* $A_{\beta,n}$ (over K), with many interesting combinatorial properties. For its basic structure see F. M. Goodman, P. de la Harpe, and V. F. R. Jones, *Coxeter Graphs and Towers of Algebras*, Springer-Verlag, New York, 1989, p. 33 and §2.8. For a direct connection with 321-avoiding permutations (defined in item 115), see S. C. Billey, W. Jockusch, and R. Stanley, *J. Algebraic Combinatorics* **2** (1993), 345–374 (pp. 360–361).

(h) See J.-Y. Shi, *Quart. J. Math.* **48** (1997), 93–105 (Thm. 3.2(a)).

(i) This remarkable result is a small part of a vast edifice due primarily to M. Haiman, *J. Algebraic Combinatorics* **3** (1994), 17–76. See also A. M. Garsia and M. Haiman, *J. Algebraic Combinatorics* **5** (1996), 191–244; A. M. Garsia and M. Haiman, *Electron. J. Combinatorics* **3** (1996), no. 2, Paper 24; and M. Haiman, *Discrete Math.* **193** (1998), 201–224, as well as Problem A44 of the present monograph.

(j) The degree of $G(k, n + k)$ is the number $f^{(n^k)}$ of standard Young tableaux of the rectangular shape (n^k) (see e.g. R. Stanley, *Lecture Notes in Math.* **579**, Springer, Berlin, 1977, pp. 217-251 (Thm. 4.1) or L. Manivel, *Symmetric Functions, Schubert Polynomials and Degeneracy Loci*, American Mathematical Society and Société Mathématique de France, 1998), and the proof follows from item 168.

(k) This result was conjectured by J.-C. Aval, F. Bergeron, N. Bergeron, and A. Garsia. The "stable case" (i.e., $n \to \infty$) was proved by J.-C. Aval and N. Bergeron, *Proc. Amer. Math. Soc.* **131** (2002), 1053–1062. The full conjecture was proved by J.-C. Aval, F. Bergeron, and N. Bergeron, *Advances in Math.* **181** (2004), 353–367.

(l) This is a result of Alexander Woo, math.CO/0407160. Woo conjectures that Ω_w is the "most singular" Schubert variety, i.e., the point X_{w_0} (which always has the largest multiplicity for any Schubert variety Ω_v) of Ω_w has the largest multiplicity of any point on any Schubert variety of $GL(n, \mathbb{C})/B$.

(m) A matrix $A \in SL(n, \mathbb{C})$ satisfying $A^{n+1} = 1$ is diagonalizable with eigenvalues ζ satisfying $\zeta^{n+1} = 1$. The conjugacy class of A is then determined by its multiset of eigenvalues. It follows that the number of conjugacy classes is the number of multisets on $\mathbb{Z}/(n + 1)\mathbb{Z}$ whose elements sum to 0. Now use item 192. For the significance of this result and its generalization to other Lie groups, see D. Z. Djoković, *Proc. Amer. Math. Soc.* **80** (1980), 181–184, and T. Friedmann and R. Stanley, *Europ. J. Combinatorics* **36** (2014), 86–96. Further discussion appears in Lecture 5 of S. Fomin and N. Reading, in *Geometric Combinatorics*, IAS/Park City Math. Ser. **13**, American Mathematical Society, Providence, RI, 2007, pp. 63–131.

(n) The determinant of a square submatrix B of A will not vanish if and only if the main diagonal entries of B are nonzero. Numerous simple bijections are now available to complete the proof.

(o) This result can be derived from D. G. Mead, *Pacific J. Math.* **42** (1972), 165–175 (proof of Lemma 1); J. Désarménien, J.P.S. Kung, and G.-C. Rota, *Advances in Math.* **27** (1978), 63–92 (proof of Theorem 2.2); and P. Doublilet, G.-C. Rota, and J. Stein, *Studies in Applied Math.* **53** (1974), 185–216 (proof of Theorem 1). The relevant idea is more explicit in J. Désarménien, *Discrete Math.* **30** (1980), 51–68 (Theorem 2). An explicit statement and proof appears in B. Rhoades and M. Skandera, *Ann. Combinatorics* **9** (2005), 451–495 (Proposition 4.7).

(p) By definition of matrix multiplication, we have

$$(A^{2n})_{11} = \sum a_{1,i_1} a_{i_1,i_2} \cdots a_{i_{2n-1},1}.$$

Each nonzero term in this sum is equal to 1. There is a bijection between such terms and the ballot sequence of item 77, given by associating the term $a_{1,i_1} a_{i_1,i_2} \cdots a_{i_{2n-1},1}$ with the ballot sequence

$$i_1 - 1, i_2 - i_1, i_3 - i_2, \ldots, 1 - i_{2n-1}.$$

Another result along these lines is that if $B = (b_{ij})_{i,j \geq 1}$ is given by $b_{ij} = 1$ for $j - i \geq -1$ and otherwise $b_{ij} = 0$, then $(B^n)_{ij} = C_n$. For a further variant, see the solution to item 51. For a vast generalization, see P. Monsky, *Electronic J. Combinatorics* **18**(1) (2011), P5. An interesting related paper is A. Berenstein, V. Retakh, C. Reutenauer, and D. Zeilberger, in *Contemporary Math.* **592** (2013), pp. 103–110.

(q) See T. Fonseca and P. Zinn-Justin, *Electr. J. Comb.* **15** (2008), #R81 (Appendix C). A later reference is D. Romik, arXiv:1303.6341. The polynomials under consideration are called *wheel polynomials* and have many fascinating properties connected with noncrossing matchings, loop percolation, the quantum Knizhnik-Zamolodchikov equation, and other topics.

A9. (a) These results appear in M. Bousquet-Mélou and G. Schaeffer, *Probab. Theory Related Fields* **124** (2002), 305–344 (Theorem 7), and M. Bousquet-Mélou, *Adv. in Appl. Math.*, **27** (2001), 243–288 (Theorem 19). The proofs are obtained from the formula

$$\sum_{n \geq 0} \sum_{i,j} a_{i,j}(n) x^i y^j t^n =$$

$$\frac{\left(1 - 2t(1+\bar{x}) + \sqrt{1-4t}\right)^{1/2} \left(1 + 2t(1-\bar{x}) + \sqrt{1+4t}\right)^{1/2}}{2(1 - t(x+\bar{x}+y+\bar{y}))},$$

(5.1)

where $\bar{x} = 1/x$ and $\bar{y} = 1/y$. Equation (4.1) is also given a bijective proof in the second paper (Proposition 2).

(b) Equation (4.2) was conjectured by Bousquet-Mélou and Schaeffer, *Probab. Theory Related Fields* **124** (2002) §3.1. This conjecture, as well as Equation (4.3), was proved by G. Xin, *Discrete Math.* **282** (2004), 281–287. The proof is obtained from (5.1) as in (a).

(c) The situation is analogous to (a). The results appear in the two papers cited in (a) and are based on the formula

$$\sum_{n \geq 0} \sum_{i,j} b_{i,j}(n) x^i y^j t^n = \frac{\left(1 - 8t^2(1+\bar{x}^2) + \sqrt{1-16t^2}\right)^{1/2}}{\sqrt{2}(1 - t(x+\bar{x})(y+\bar{y}))}.$$

The case $i = 1$ of (4.4) is given a bijective proof in Bousquet-Mélou and Schaeffer, *Probab. Theory Related Fields* **124** (2002), Proposition 7.

(d) Let X_n be the set of all closed paths of length $2n$ from $(0,0)$ to $(0,0)$ that intersect the half-line L defined by $y = x$, $x \geq 1$. Given $P \in X_n$, let k be the smallest integer such that P intersects L after k steps, and let Q be the path consisting of the first k steps of P. Let P' be the path obtained from P by reflecting Q about the line $y = x$. Then $P' \in X_n$, $w(P') = w(P) \pm 1$, and the map $P \mapsto P'$ is an involution. Any path P of length $2n$ from $(0,0)$ to $(0,0)$ not contained in X_n satisfies $w(P) = 0$. It follows that $f(n)$ is the number of closed paths of length $2n$ from $(0,0)$ to $(0,0)$ not intersecting L.

Now consider the linear change of coordinates $(x,y) \mapsto \frac{1}{2}(-x - y, x - y)$. This transforms a closed path of length $2n$ from $(0,0)$ to $(0,0)$ not intersecting L to a closed path from $(0,0)$ to $(0,0)$ with $2n$ diagonal steps $(\pm 1, \pm 1)$, not intersecting the negative real axis. Now use Equation (4.5).

This result was stated by R. Stanley, Problem 10905, Problems and Solutions, *Amer. Math. Monthly* **108** (2001), 871. The published solution by R. Chapman, **110** (2003), 640–642, includes a self-contained proof of Equation (4.5).

A10. Since every binary tree with at least one edge has a vertex of odd degree (e.g., an endpoint), $s(T)$ is half the number of vertices of T of odd degree by a basic result in graph theory. For $i = 1, 2$ let $f_i(k, n)$ be the number of binary trees on n vertices with $2k$ vertices of odd degree such that the degree of the root has the same parity as i. Set

$$F_i = F_i(x, y) = \sum_{k,n} f_i(k, n) x^k y^n.$$

It is easy to see that

$$F_1 = 2xF_1 + 2xyF_2$$
$$F_2 = x + xy^{-1}F_1^2 + 2xF_1F_2 + xyF_2^2.$$

One can solve for F_1 and F_2 and compute that

$$\sum_{n \geq 1} t(n) C_n x^n = \frac{\partial}{\partial y}(F_1 + F_2)|_{y=1}$$

$$= \frac{1 - 6x + 10x^2}{2x\sqrt{1 - 4x}} - \frac{1}{2x}(1 - 4x + 4x^2),$$

from which the stated result is a routine computation.

A11. (a) The enumeration of rooted planar maps subject to various conditions is a vast subject initiated by W. T. Tutte. The result of this problem appears in *Canad. J. Math.* **15** (1963), 249–271. A good introduction to the subject, with many additional references, can be found in Goulden and Jackson [26, §2.9]. There has been a revival of interest in the enumeration of maps, motivated in part by connections with physics. At present there is no comprehensive survey of this work, but two references that should help combinatorialists get into the subject are D. M. Jackson, *Trans. Amer. Math. Soc.* **344** (1994), 755–772; and D. M. Jackson, in *DIMACS Series in Discrete Mathematics and Computer Science* **24** (1996), 217–234.

(b) Let G be a rooted planar map with n edges and p vertices. We can define a dual n-edge and p'-vertex map G', such that by Euler's formula $p' = n + 2 - p$. From this it follows easily that the answer is given by $M(n)(n+2)/2$, where $M(n)$ is the answer to (a). For a more general result of this nature, see V. A. Liskovets, *J. Combinatorial*

Theory, Ser. B **75** (1999), 116–133 (Prop. 2.6). It is also mentioned on page 150 of L. M. Koganov, V. A. Liskovets, and T.R.S. Walsh, *Ars Combinatoria* **54** (2000), 149–160.

(c) This result was first proved nonbjectively by R. C. Mullin, *Canad. J. Math.* **19** (1967), 174–183. The first bijective proof was given by R. Cori, S. Dulucq, and G. Viennot, *J. Combinatorial Theory, Ser. A* **43** (1986), 1–22. A nicer bijective proof was given by O. Bernardi, *Electr. J. Comb.* **124** (2007), #R9.

A12. (a) See T. Nakamigawa, *Theoretical Computer Science* **235** (2000), 271–282 (Corollary 6); and A. Dress, J. Koolen, and V. Moulton, *European J. Combin.* **23** (2002), 549–557. Another proof was given by J. Jonsson in the reference cited below.

(b) See J. Jonsson, *J. Combinatorial Theory, Ser. A* **112** (2005), 117–142. The proof interprets the result in terms of lattice path enumeration and applies the Gessel-Viennot theory of nonintersecting lattice paths [64, §2.7].

(c) See L. Serrano and C. Stump, *Electr. J. Comb.* **19** (2012), P16.

A13. This result was first proved by I. Dumitriu using random matrix arguments. An elegant bijective proof was then given by E. Rassart, as follows. We want a bijection φ from (1) quadruples (P, P', e, e'), where P and P' are Dyck paths with $2n$ steps and e and e' are up edges of P and P' ending at the same height i; and (2) Dyck paths Q with $4n$ steps from $(0,0)$ to $(4n,0)$ which do not touch the point $(2n,0)$. We first transform (P, e) into a partial Dyck path L ending at height i. Let f be the down edge paired with e (i.e., the first down edge after e beginning at height i), and flip the direction of each edge at or after f. Let e_2 be the first up edge to the left of e ending at height $i - 1$, and let f_2 be the down edge paired with e_2 in the original path P. Flip all edges of the current path at or after f_2. Continue this procedure, letting e_j be the first edge to the left of e_{j-1} such that e_j has height one less than e_{j-1}, etc., until no edges remain. We obtain the desired partial Dyck path L ending at height i. Do the same for (P', e'), obtaining another partial Dyck path L' ending at height i. Reverse the direction of L' and glue it to the end of L. This gives the Dyck path Q. Figure 5.1 shows an example of the correspondence $(P, e) \mapsto L$. We leave the construction of φ^{-1} to the reader. See I. Dumitriu and E. Rassart, *Electr. J. Comb.* **10**(1) (2003), R-43.

A14. Fuss-Catalan numbers first appeared in a paper of N. Fuss, *Nova Acta Academiae Sci. Petropolitanae* **9** (1791), 243–251, in response to a question of J. Pfaff. For this reason Fuss-Catalan numbers are sometimes called *Pfaff-Fuss-Catalan numbers*. Sometimes the more

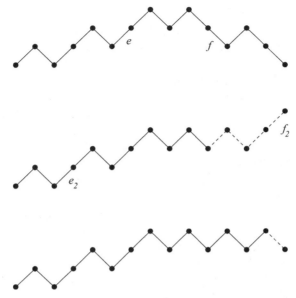

Figure 5.1. The bijection $(P, e) \mapsto L$ in the solution to Problem A13.

general numbers $\frac{r}{kn+r}\binom{(k+1)n+r-1}{n}$ are called Fuss-Catalan numbers. Parts (a)–(d) are straightforward extensions of the Catalan ($k = 1$) case. For (e) see P. H. Edelman, *Discrete Math.* **31** (1980), 171–180; [64, Exercise 3.158]; and D. Armstrong, *Memoirs Amer. Math. Soc.* **202** (2009), no. 949. For (f), see S. Yakoubov, arXiv:1310.2979. For a more extensive discussion of Fuss-Catalan numbers, see S. Heubach, N. Y. Li, and T. Mansour, A garden of k-Catalan structures, preprint.

A15. Let w be the vertex of F adjacent to v. Let G denote F with v removed, and let $g(n)$ be the number of closed walks in G of length $2n$ beginning at w. Write $A(x) = \sum_{n \geq 0} f(n) x^n$ and $B(x) = \sum_{n \geq 0} g(n) x^n$. The definitions of F and G yield

$$A(x) = 1 + xA(x)B(x)$$
$$B(x) = 1 + xB(x)^2 + xA(x)B(x).$$

Eliminating $B(x)$ gives $A(x) = 1 + xA(x)^3$, the equation satisfied by the generating function for ternary trees by number of vertices. This result first appeared in D. Bisch and V. Jones, *Invent. math.* **128** (1997), 89–157; and D. Bisch and V. Jones, in *Geometry and Physics* (Aarhus, 1995), Lecture Notes in Pure and Appl. Math., vol. **184**, Dekker, New York, 1997. See [65, Exercises 5.45, 5.46, 5.47(b)] for further occurrences of the *ternary Catalan number* $\frac{1}{2n+1}\binom{3n}{n}$ (the case $k = 2$ of the Fuss-Catalan numbers of Problem A14).

A16. (b) The proof given in Section 1.6 (the case $(a,b) = (n,\ n+1)$) generalizes straightforwardly. The connection between (a,b)-Catalan numbers and lattice paths was first stated by H. D. Grossman, *Duke Math. J.* **14** (1950), 305–313. The first published proof is due to M.T.L. Bizley, *J. Inst. Actuaries* **80** (1954), 55–62. D. Armstrong has collected and discovered many additional aspects of (a,b)-Catalan numbers. See for instance www.math.miami.edu/~armstrong/Talks/RCC_CAAC.pdf.

A17. E. Catalan stated in *Nouvelles Annales de Mathématiques: Journal des Candidats aux École Polytechnic et Normale*, Series 2, **13** (1874), 207, that $S(m,n)$ is an integer. I. Gessel, *J. Symbolic Computation* **14** (1992), 179–194 (Section 6), revived interest in the topic and raised the question of a combinatorial interpretation. Some special cases are known, but the problem remains open in general.

A18. See S. Choi and H. Park, *J. Math. Soc. Japan*, to appear, arXiv:1210.3776 (Theorem 2.5) for the case $k = 2$.

A19. This result (in a much more general context) is due to B. Drake, PhD thesis, Brandeis University, 2008 (Example 1.9.3). A bijective proof was found by S.J.X. Hou, *J. Xinjiang University (Natural Science Edition)*, issue 2 (2013), 165–169.

A20. Private communication from V. Tewari, February 18, 2011. For some related results, see V. Tewari, arXiv:1403.5327.

A21. (a) The most straightforward method is to observe that all three power series $F(t_0, t_1, \ldots; x)$ satisfy

$$F(t_0, t_1, \ldots; x) = \frac{1}{1 - t_0 x F(t_1, t_2, \ldots; x)}$$

with the initial condition

$$F(t_0, t_1, \ldots; 0) = 1.$$

A general combinatorial theory of continued fractions is due to P. Flajolet, *Discrete Math.* **32** (1980), 125–161. The equivalence of (i) and (iii) is equivalent to a special case of Corollary 2 of this paper.

(b) Put $t_0 = t_1 = \cdots = t_n = 1$ and $t_i = 0$ for $i > n$. By the case $j = 2$ of [64, Exercise 1.34], the generating function of part (ii) above becomes the left-hand side of Equation (4.7). On the other hand, by item 6 the coefficient of x^i for $i \leq n + 1$ in the generating function of part (iii) becomes the coefficient of x^i in the right-hand side of (4.7) (since a tree with $i \le n+1$ vertices has height at most n). Now let $n \to \infty$. It is not difficult to give a direct proof of (4.7). The generating function $F_n(x) = \sum_{i \ge 0} (-1)^i \binom{n-i}{i} x^i$ satisfies

$$F_{n+2}(x) = F_{n+1}(x) - xF_n(x). \tag{5.2}$$

From [64, Thm. 4.1.1] it follows that

$$F_n(x) = \frac{C(x)^{-n-1} - (xC(x))^{n+1}}{\sqrt{1-4x}}.$$

Hence the fraction in the left-hand side of (4.7) is given by

$$\frac{\sum_{i \ge 0} (-1)^i \binom{n-i}{i} x^i}{\sum_{i \ge 0} (-1)^i \binom{n+1-i}{i} x^i} = C(x) \frac{1 - x^{n+1} C(x)^{2(n+1)}}{1 - x^{n+2} C(x)^{2(n+2)}},$$

and (4.7) follows.

Equation (4.7) (with the numerator and denominator on the left-hand side defined by (5.2)) was given by V. E. Hoggatt, Jr., Problem H-297, *Fibonacci Quart.* **17** (1979), 94; solution by P. S. Bruckman, **18** (1980), 378.

(c) This result is due to A. Postnikov and B. Sagan, *J. Combinatorial Theory, Ser. A* **114** (2007), 970–977. It was originally conjectured by Postnikov for the case $t_i = (2i+1)^2$. A q-analogue is due to M. Konvalinka, *J. Combinatorial Theory, Ser. A* **114** (2007), 1089–1100.

A22. These are unpublished results of A. Postnikov and R. Stanley. Part (d) is a consequence of Problem A24. For proofs, variations, and strengthenings, see K. Mészáros, *Trans. Amer. Math. Soc.* **363** (2011), 4359–4382, 6111–6141.

A23. This result was conjectured by F. Brenti in 1995 and first proved by D. Zeilberger,

www.math.rutgers.edu/~zeilberg/mamarim/mamarimhtml/
catalan.html.

Zeilberger's proof consists essentially of the statement

$$B_{2n}(q) = (-1)^n \frac{1}{2n+1} \binom{2n+1}{n} q^n$$

$$+ \sum_{i=0}^{n-1} (-1)^i \frac{2n-2i+1}{2n+1} \binom{2n+1}{i} (q^i + q^{2n-i})$$

$$B_{2n+1}(q) = \sum_{i=0}^{n} (-1)^{i+1} \frac{n-i+1}{n+1} \binom{2n+2}{i} q^i.$$

Later (as mentioned by Zeilberger, ibid.) Brenti found a combinatorial interpretation of the polynomials $B_m(q)$ which implies his conjecture.

A24. (a,c) Let M' denote the part of M below the main diagonal. It is easy to see that M' uniquely determines M. It was shown by C. S. Chan, D. P. Robbins, and D. S. Yuen, *Experiment. Math.* **9** (2000), 91–99, that the polytope CR_{n+1} (called the *Chan-Robbins-Yuen polytope* or *CRY polytope*) can be subdivided into simplices, each of relative volume $1/\binom{n+1}{2}!$, which are naturally indexed by the matrices M'. It follows that $\binom{n+1}{2}! \nu(CR_{n+1}) = g(n)$. Chan and Robbins had earlier conjectured that $\binom{n+1}{2}! \nu(CR_{n+1}) = C_1 C_2 \cdots C_n$. (Actually, Chan and Robbins use a different normalization of relative volume so that $\nu(CR_{n+1}) = C_1 C_2 \cdots C_n$.) This conjecture was proved by D. Zeilberger, *Electron. Trans. Numer. Anal.* **9** (1999), 147–148, and later W. Baldoni-Silva and M. Vergne, math.CO/0103097 (Thm. 33).

(b) Given the matrix $M = (m_{ij})$, there exist unique nonnegative integers $a_1, \ldots, a_n$ satisfying

$$\sum_{1 \le i < j \le n} m_{ji}(e_i - e_j) + \sum_{i=1}^{n} a_i(e_i - e_{n+1}) = \left(1, 2, \ldots, n, -\binom{n+1}{2}\right).$$

This sets up a bijection between (a) and (b). This result is due to A. Postnikov and R. Stanley (unpublished), and is also discussed by W. Baldoni-Silva and M. Vergne, math.CO/0103097 (§8).

A25. Let the vertices be $1, 2, \ldots, 4m+2$ in clockwise order. Suppose that there is a chord between vertex 1 and vertex $2i$.

Case 1: i is odd. The polygon is divided into two polygons, one (say P_1) with vertices $2, \ldots, 2i - 1$ and the other (say P_2) with vertices $2i + 1, \ldots, 4m + 2$. Choose $i - 1$ noncrossing chords on the vertices $2, 3, \ldots, 2i - 1$. Thus we have a net N_1 on P_1. Choose a coloring of the faces of N_1. By "symmetry" half the nets on P_2 will have an even number of blue faces and half an odd number (with respect to the coloring of the faces of N_1), so the number of nets with a chord $(1, 2i)$ and an even number of blue faces minus the number of nets with a chord $(1, 2i)$ and an odd number of blue faces is 0. (It's easy to make this argument completely precise.)

Case 2: i is even, say $i = 2j$. The number of nets N_1 on P_1 with an even number of blue faces is $f_e(j - 1)$. The number with an odd number is $f_o(j - 1)$. Similarly, the number of nets N_2 on P_2 with an even number of blue faces is $f_e(m - j)$ and with an odd number of blue faces is $f_o(m - j)$. Hence the number of nets on the original $4m + 2$ points with a chord $(1, 4j)$ and an even number of blue faces is $f_e(j - 1)f_e(m - j) + f_o(j - 1)f_o(m - j)$. Thus

$$f_e(m) = \sum_{j=1}^{m} (f_e(j - 1)f_e(m - j) + f_o(j - 1)f_o(m - j)).$$

Similarly

$$f_o(m) = \sum_{j=1}^{m} (f_e(j - 1)f_o(m - j) + f_o(j - 1)f_e(m - j)).$$

Hence

$$f_e(m) - f_o(m) = \sum_{j=1}^{m} (f_e(j - 1) - f_o(j - 1))(f_e(m - j) - f_o(m - j)),$$

which is just the recurrence satisfied by Catalan numbers C_m (initial condition $C_0 = 1$) and thus also by $D_m = (-1)^{m+1} C_m$ (initial condition $D_0 = -1$). Since $f_e(0) - f_o(0) = 0 - 1 = -1$, the proof follows.

This result was first proved using techniques from algebraic geometry by A. Eremenko and A. Gabrielov, in *Computational Methods and Function Theory*, vol. 1, 2001, pp. 1–25. A bijective proof was given by S.-P. Eu, Coloring the net (private communication), based on the paper S.-P. Eu, S.-C. Liu, and Y.-N. Yeh, *Discrete Math.* **281** (2004), 189–196.

A26. This result is due to G. Berkolaiko, J. Harrison, and M. Novaes, *J. Phys. A: Math. Theor.* **41** (2008), #365102. The result was originally suggested to Novaes by some calculations in random matrix theory. Subsequently Novaes and his collaborators found a generating function proof.

A27. A conjectured affirmative answer (to a slightly different statement of the problem) is due to J. Körner, C. Malvenuto, and G. Simonyi, *SIAM J. Discrete Math.* **22**(2) (2008), 489–499 (§3.2). These authors show by a construction that $f(n) \geq C_n$. Jineon Baek gave the following counterexample for $n = 5$ (private communication August 9, 2012): {(1,2,4,8,16), (2,3,4,8,16), (1,4,5,8,16), (4,5,6,8,16), (4,6,7,8,16), (1,2,8,9,16), (2,3,8,9,16), (1,8,9,10,16), (1,8,10,11,16), (8,9,10,12,16), (8,10,11,12,16), (8,9,12,13,16), (8,12,13,14,16), (8,12,14,15,16), (1,2,4, 16,17), (2,3,4,16,17), (1,4,5,16,17), (4,5,6,16,17), (4,6,7,16,17), (1,2,16, 17,18), (2,3,16,20,21), (3,4,16,18,19), (4,5,16,18,20), (2,4,16,19,20), (1,16,17,18,20), (1,16,18,19,20), (1,16,17,20,21), (1,16,20,21,22), (1,16, 20,22,23), (16,17,18,20,24), (16,18,19,20,24), (16,17,20,21,24), (16,20, 21,22,24), (16,20,22,23,24), (16,17,18,24,25), (16,18,19,24,25), (16,17, 24,25,26), (16,17,24,26,27), (16,24,25,26,28), (16,24,26,27,28), (16,24, 25,28,29), (16,24,28,29,30), (16,24,28,30,31)}. This set has 43 elements. NOTE. This problem has been included to show that, sadly, Catalan numbers are not the solution to every enumeration problem! The next problem is of a similar nature.

A28. This conjecture is due to S. Garrabrant and I. Pak, `arXiv:1407.8222` (Conjecture 4.5). The authors show that the conjecture is equivalent to a conjecture about tiling a strip with tiles of possibly irrational length.

A29. (a) The techniques of R. Stanley, *J. Combinatorial Theory, Ser. A* **114** (2007), 436–460, can be used to show that the coefficients in question count 321-avoiding alternating permutations in $\mathfrak{S}_{2m-1}$. Now use item 146.

    (b) Similar to (a), we are counting 321-avoiding alternating permutations in $\mathfrak{S}_{2m}$. Again use item 146.

    (c) Here we are counting 321-avoiding reverse alternating permutations in $\mathfrak{S}_{2m}$. Now use item 147.

A30. From Equation (1.4) or otherwise, we have $A(x)^2 - A(x) + x = 0$ and $x(B(x) + 1)^2 - B(x) = 0$. Hence $x = A(x) - A(x)^2$ and $x = B(x)/(1 + B(x)^2)$, so

$$A(x)^{\langle -1 \rangle} = x - x^2$$
$$B(x)^{\langle -1 \rangle} = \frac{x}{(1+x)^2}.$$

A31. We have $1 + x\frac{d}{dx}\log(1 + y) = 1/(1 - y)$. It is routine to solve this differential equation, yielding $y = C(\alpha x)$, where α is a constant.

A32. By polynomial interpolation it suffices to prove the two identities when q is a positive integer.

(a) Let $G(x) = xC(x)$. Then $G(x) = (x - x^2)^{\langle -1 \rangle}$, and the proof follows easily from the Lagrange inversion formula [65, Thm. 5.4.2]. Alternatively, $C(x)^q$ (when $q \in \mathbb{P}$) is the generating function for plane binary forests with q components. Now use [65, Thm. 5.3.10] in the case $n = 2k + q$, $r_0 = k + q$, $r_2 = k$ (and all other $r_i = 0$).

(b) This can be obtained from (a) by differentiating $(xC(x))^q$ with respect to x, or alternatively from the identity

$$xC(x)^q = \frac{C(x)^{q-1}}{2} - \frac{(1-4x)C(x)^{q-1}}{2\sqrt{1-4x}}.$$

A33. (a) It is straightforward to verify that $\left(\sum_{n \geq 0} C_{2n} x^{2n}\right)^2 = \sum 4^n C_n x^{2n}$, which is equivalent to (4.9). This result is due to L. Shapiro, private communication dated May 24, 2002.

(b) See G. E. Andrews, *Adv. in Appl. Math.* **46** (2011), 15–24; and G. V. Nagy, *Adv. in Appl. Math.* **49** (2012), 391–396.

A34. (a) It is easy to see that if $F(t)$ exists, then it is unique. Now it follows from $tC(t)^2 - C(t) + 1 = 0$ that

$$\frac{1 - x + xtC(t)}{1 - x + x^2 t} = \frac{1}{1 - xtC(t)}.$$

Hence $F(t) = C(t)$. For an application, see item 23.

(b) We have

$$\frac{1}{1 - xtC(t)} = \sum_{k \geq 0} x^k t^k C(t)^k.$$

Hence by (a) and Problem A32(a) there follows

$$[x^k] f_n(x) = [t^n] t^k C(t)^k$$
$$= [t^{n-k}] C(t)^k$$
$$= \frac{k}{n} \binom{2n - k - 1}{n - k} = \frac{k}{n} \binom{2n - k - 1}{n - 1}.$$

A35. (a) This curious result can be proved by induction using suitable row and column operations. Alternatively, it is a straightforward application of the nonintersecting lattice path method of evaluating determinants [64, Thm. 2.7.1]. It arose from a problem posed by E. Berlekamp which was solved (in a stronger form) by L. Carlitz, D. P. Roselle, and R. A. Scoville, *J. Combinatorial Theory* **11** (1971), 258–271. A slightly different way of stating the result appears in R. Stanley, *Fib. Quart.* **13** (1975), 215–232 (page 223). A generalization is due to C. Bessenrodt and R. Stanley, *J. Alg. Comb.* to appear; arXiv:1311.6123. For a special case of this generalization, see Problem A43(b).

(b) *Answer:* $a_n = C_n$, the nth Catalan number. One way (of many) to prove this result is to apply part (a) to the cases $\lambda = (2n + 1, 2n, \ldots, 2, 1)$ and $\lambda = (2n, 2n - 1, \ldots, 2, 1)$, and to use the interpretation of Catalan numbers given by item 24. Related work appears in A. Kellogg (proposer), Problem 10585, *Amer. Math. Monthly* **104** (1997), 361; and C. Radoux, *Bull. Belgian Math. Soc. (Simon Stevin)* **4** (1997), 289–292.

A36. (a) The unique such basis $y_0, y_1, \ldots, y_n$, up to sign and order, is given by

$$y_j = \sum_{i=0}^{j} (-1)^{j-i} \binom{i+j}{2i} x_i.$$

(b) Now

$$y_j = \sum_{i=0}^{j} (-1)^{j-i} \binom{i+j+1}{2i+1} x_i.$$

A37. (a) The problem of computing the probability of convexity was raised by J. van de Lune and solved by R. B. Eggleton and R. K. Guy, *Math. Mag.* **61** (1988), 211–219, by a clever integration argument. The proof of Eggleton and Guy can be "combinatorialized" so that integration is avoided. The decomposition of $\mathcal{C}_d$ given below in the solution to (c) also yields a proof. For a more general result, see P. Valtr, in *Intuitive Geometry* (Budapest, 1995), *Bolyai Soc. Math. Stud.* **6**, János Bolyai Math. Soc., Budapest, 1997, pp. 441–443.

(b) Suppose that $x = (x_1, x_2, \ldots, x_d) \in \mathcal{C}_d$. We say that an index i is *slack* if $2 \leq i \leq d - 1$ and $x_{i-1} + x_{i+1} > 2x_i$. If no index is slack, then either $x = (0, 0, \ldots, 0)$, $x = (1, 1, \ldots, 1)$, or $x = \lambda(1, 1, \ldots, 1) + (1 - \lambda)y$ for $y \in \mathcal{C}_d$ and sufficiently small $\lambda > 0$. Hence in this last case x is not a vertex. So suppose that x has a slack index. If for all slack indices i we have $x_i = 0$, then x is of the stated form (4.10). Otherwise, let i be a slack index such that $x_i > 0$. Let $j = i - p$ be the largest index such that $j < i$ and j is not slack. Similarly, let $k = i + q$ be the smallest index such that $k > i$ and k is not slack. Let

$$A(\epsilon) = \left(x_1,\ldots,x_j,x_{j+1}+\frac{\epsilon}{p},x_{j+2}+\frac{2\epsilon}{p},\ldots,\right.$$
$$\left. x_i+\epsilon,\ldots,x_{k-2}+\frac{2\epsilon}{q},x_{k-1}+\frac{\epsilon}{q},x_k,\ldots,x_n\right).$$

For small $\epsilon > 0$, both $A(\epsilon)$ and $A(-\epsilon)$ are in C_d. Since $x = \frac{1}{2}\left(A(\epsilon)+\frac{1}{2}A(-\epsilon)\right)$, it follows that x is not a vertex. The main idea of this argument is due to A. Postnikov.

(c) For $1 \leq r \leq s \leq d$, let

$$F_{rs} = \{(x_1,\ldots,x_d) \in C_d : x_r = x_{r+1} = \cdots = x_s = 0\}$$
$$F_r^- = \{(x_1,\ldots,x_d) \in C_d : x_r = x_{r+1} = \cdots = x_d = 0\}$$
$$F_s^+ = \{(x_1,\ldots,x_d) \in C_d : x_1 = x_2 = \cdots = x_s = 0\}.$$

Now F_r^- is a simplex with vertices $(1, \frac{k-1}{k}, \frac{k-2}{k}, \ldots, \frac{1}{k}, 0, 0, \ldots, 0)$ for $1 \leq k \leq r-1$, together with $(0,0,\ldots,0)$. It can then be shown that

$$\sum_{n\geq 0} i(F_r^-,n)x^n = \frac{1}{[1][r-1]!}.$$

Similarly

$$\sum_{n\geq 0} i(F_s^+,n)x^n = \frac{1}{[1][d-s]!}.$$

Since $F_{rs} \cong F_r^- \times F_s^+$, we have $i(F_{rs},n) = i(F_r^-,n)i(F_s^+,n)$ and

$$\sum_{n\geq 0} i(F_{rs},n)x^n = \frac{1}{[1][r]!} * \frac{1}{[1][d-s]!}.$$

Let P be the poset of all F_{rs}'s, ordered by inclusion, and let μ denote the Möbius function of $P \cup \{\hat{1}\}$. Let $G = \bigcup_{r=1}^d F_{rr}$, a polyhedral complex in $\mathbb{R}^d$. By Möbius inversion we have

$$i(G,n) = -\sum_{F_{st}\in P} \mu(F_{st},\hat{1})i(F_{st},n).$$

But $F_{tu} \subseteq F_{rs}$ if and only if $t \leq r \leq s \leq u$, from which it is immediate that

$$-\mu(F_{st},\hat{1}) = \begin{cases} 1, & s=t \\ -1, & s=t-1 \\ 0, & \text{otherwise}. \end{cases}$$

Hence

$$\sum_{n\geq 0} i(G,n)x^n = \sum_{r=1}^{d} \frac{1}{[1][r-1]!} * \frac{1}{[1][d-r]!}$$

$$-\sum_{r=1}^{d-1} \frac{1}{[1][r-1]!} * \frac{1}{[1][d-1-r]!}.$$

Now the entire polytope $\mathcal{C}_d$ is just a cone over G with apex $(1,1,\ldots,1)$. From this it is not hard to deduce that

$$\sum_{n\geq 0} i(\mathcal{C}_d,n)x^n = \frac{1}{1-x}\sum_{n\geq 0} i(G,n)x^n,$$

and the proof follows.

A38. See P. Valtr, *Discrete Comput. Geom.* **13** (1995), 637–643. Valtr also shows in *Combinatorica* **16** (1996), 567–573, that if n points are chosen uniformly and independently from inside a triangle, then the probability that the points are in convex position is $\frac{2^n}{(2n)!}\binom{3n-3}{n-1,n-1,n-1}$.

A39. Equation (4.12) is equivalent to

$$\sum_{n\geq 0} f_n \frac{1}{n!} \left(\log(1-x)^{-1}\right)^n x^n = C(x).$$

Hence by [65, (5.25)] we need to show that

$$n!\,C_n = \sum_{k=1}^{n} c(n,k)f_k,$$

where $c(n,k)$ is the number of permutations $w \in \mathfrak{S}_n$ with k cycles. Choose a permutation $w \in \mathfrak{S}_n$ with k cycles in $c(n,k)$ ways. Let the cycles of w be the elements of a semiorder P in f_k ways. For each cycle $(a_1,\ldots,a_i)$ of w, replace this element of P with an antichain whose elements are labeled $a_1,\ldots,a_i$. If $a = (a_1,\ldots,a_i)$ and $b = (b_1,\ldots,b_j)$ are two cycles of w, then define $a_r < b_s$ if and only if $a < b$ in P. In this way we get a poset $\rho(P,w)$ with vertices $1,2,\ldots,n$. It is not hard to see that $\rho(P,w)$ is a semiorder, and that every isomorphism class of n-element semiorders occurs exactly $n!$ times among the posets $\rho(P,w)$. Since by item 180 there are C_n nonisomorphic n-element semiorders, the proof follows.

This result was first proved by J. L. Chandon, J. Lemaire, and J. Pouget, *Math. Sci. Hum.* **62** (1978), 61–80, 83. For a more general situation in which the number A_n of unlabeled objects is related to the number B_n of labeled objects by $\sum B_n x^n/n! = \sum A_n(1 - e^{-x})^n$, see R. Stanley, *Proc. Natl. Acad. Sci. U.S.A.* **93** (1996), 2620–2625 (Thm. 2.3); A. Postnikov and R. Stanley, *J. Combinatorial Theory Ser. A* **91** (2000), 544–597 (§6); and [64, Exercise 3.17] (due to Y. Zhang).

A40. (a) (sketch) We will triangulate $\mathcal{P}$ into d-dimensional simplices σ, all containing 0. Thus each σ will have d vertices of the form $e_i - e_j$, where $i < j$. Given a graph G with d edges on the vertex set $[d+1]$, let σ_G be the convex hull of all vectors $e_i - e_j$ for which ij is an edge of G with $i < j$, and let $\tilde{\sigma}_G$ be the convex hull of σ_G and the origin. It is easy to see that $\tilde{\sigma}_G$ is a d-dimensional simplex if and only if G is a tree. Moreover, it can be shown that σ_G lies on the boundary of $\mathcal{P}$ (and hence can be part of a triangulation of the type we are looking for) if and only if G is an *alternating tree*, as defined in [65, Exercise 5.41]. We therefore want to choose a set $\mathcal{T}$ of alternating trees T on $[d+1]$ such that the $\tilde{\sigma}_T$'s are the facets of a triangulation of $\mathcal{P}$. One way to do this is to take $\mathcal{T}$ to consist of the *noncrossing* alternating trees on $[d + 1]$, i.e., alternating trees such that if $i < j < k < l$, then not both ik and jl are edges. By item 62 the number of such trees is C_d. (We can also take $\mathcal{T}$ to consist of alternating trees on $[d + 1]$ such that if $i < j < k < l$ then not both il and jk are edges. By item 63 the number of such trees is again C_d.) Moreover, it is easy to see that for any tree T on $[d + 1]$ we have $V(\tilde{\sigma}_T) = 1/d!$, where V denotes relative volume. Hence $V(\mathcal{P}) = C_d/d!$. This result appears in I. M. Gelfand, M. I. Graev, and A. Postnikov, in *The Arnold-Gelfand Mathematical Seminars*, Birkhäuser, Boston, 1997 (Theorem 2.3(2)). For additional information see K. Mészáros, *Trans. Amer. Math. Soc.* **363** (2011), 4359–4382, 6111–6141.

(b) Order the $\binom{d+1}{2}$ edges ij, $1 \le i < j \le d + 1$, lexicographically, e.g., $12 < 13 < 14 < 23 < 24 < 34$. Order the C_d noncrossing alternating trees $T_1, T_2, \ldots, T_{C_d}$ lexicographically by edge set, i.e., $T_i < T_j$ if for some k the first (in lexicographic order) k edges of T_i and T_j coincide, while the $(k + 1)$st edge of T_i precedes the $(k + 1)$st edge of T_j. For instance, when $d = 3$ the ordering on the noncrossing alternating trees (denoted by their set of edges) is

$$\{12, 13, 14\},\ \{12, 14, 34\},\ \{13, 14, 23\},\ \{14, 23, 24\},\ \{14, 23, 34\}.$$

One can check that $\tilde{\sigma}_{T_i}$ intersects $\tilde{\sigma}_{T_1} \cup \cdots \cup \tilde{\sigma}_{T_{i-1}}$ in a union of $j - 1$ $(d - 1)$-dimensional faces of $\tilde{\sigma}_{T_i}$, where j is the number of vertices of T_i that are less than all their neighbors. It is not hard to see that the number of noncrossing alternating trees on $[d + 1]$ for which exactly j vertices are less than all their neighbors is just the Narayana number $N(d,j)$ of Problem A46. It follows from the techniques of R. Stanley, *Annals of Discrete Math.* **6** (1980), 333–342 (especially Theorem 1.6), that

$$(1-x)^{d+1} \sum_{n \geq 0} i(\mathcal{P},n)x^n = \sum_{j=1}^{d} N(d,j)x^{j-1}.$$

One can show that this problem is essentially equivalent to Problem A22(b). See K. Mészáros, *Trans. Amer. Math. Soc.* **363** (2011) 4359–4382.

A41. The Tamari lattice was first considered by D. Tamari, *Nieuw Arch. Wisk.* **10** (1962), 131–146, who proved it to be a lattice. A simpler proof of this result was given by S. Huang and D. Tamari, *J. Combinatorial Theory (A)* **13** (1972), 7–13. The proof sketched here follows J. M. Pallo, *Computer J.* **29** (1986), 171–175. For further properties of Tamari lattices and their generalizations, see P. H. Edelman and V. Reiner, *Mathematika* **43** (1996), 127–154; A. Björner and M. L. Wachs, *Trans. Amer. Math. Soc.* **349** (1997), 3945–3975 (§9); L.-F. Préville-Ratelle and X. Viennot, `arXiv:1406.3787`; and the references given there.

A42. (a) Since $\mathcal{S}$ is a simplicial complex, A_n is a simplicial (meet) semilattice. It is easy to see that it is graded of rank $n-2$, the rank of an element being its cardinality (number of diagonals). To check the Eulerian property, it remains to show that $\mu(S,\hat{1}) = (-1)^{\ell(S,\hat{1})}$ for all $S \in A_n$, where $\ell(S,\hat{1})$ is the length ℓ of the longest chain $S = S_0 < S_1 < \cdots < S_\ell = \hat{1}$ in the interval $[S,\hat{1}]$. If $S \in A_n$ and $S \neq \hat{1}$, then S divides the polygon C into regions $C_1,\ldots,C_j$, where each C_i is a convex n_i-gon for some n_i. Let $\bar{A}_n = A_n - \{\hat{1}\}$. It follows that the interval $[S,\hat{1}]$ is isomorphic to the product $\bar{A}_{n_1} \times \cdots \times \bar{A}_{n_j}$, with a $\hat{1}$ adjoined. It follows from [65, Exercise 5.61] (dualized) that it suffices to show that $\mu(\hat{0},\hat{1}) = (-1)^{n-2}$. Equivalently (since we have shown that every proper interval is Eulerian), we need to show that A_n has as many elements of even rank as of odd rank. One way to proceed is as follows. For any subset B of A_n, let $\eta(B)$ denote the number of elements of B of even rank minus the number of odd rank. Label the vertices of C as $1,2,\ldots,n$ in cyclic order. For $3 \leq i \leq n-1$, let $\mathcal{S}^*$ be the set of all elements of $\mathcal{S}$ for which either there is a diagonal from vertex 1 to some other vertex, or else such a diagonal can be adjoined without introducing any interior crossings. Given $S \in \mathcal{S}^*$, let i be the least vertex that is either connected to 1 by a diagonal or for which we can connect it to vertex 1 by a diagonal without introducing any interior crossings. We can pair S with the set S' obtained by deleting or adjoining the diagonal from 1 to i. This pairing (or involution) shows that $\eta(S^*) = 0$. But $A_n - S^*$ is just the interval $[T, \hat{1}]$, where T contains the single diagonal connecting 2 and n. By induction (as mentioned above) we have $\eta([T, \hat{1}]) = 0$, so in fact $\eta(A_n) = 0$.

(b) See C. W. Lee, *Europ. J. Combinatorics* **10** (1989), 551–560. An independent proof was given by M. Haiman, unpublished manuscript, 1984. This polytope is called the *associahedron*. For some far-reaching generalizations, see I. M. Gelfand, M. M. Kapranov, and A. V. Zelevinsky [24, Ch. 7], and the survey articles C. W. Lee, in *DIMACS Series in Discrete Mathematics and Theoretical Computer Science*, vol. 4, 1991, pp. 443–456, and A. Postnikov, *Int. Math. Res. Notes IMRN* (2009), 1026–1106. A nice exposition of the Gelfand-Kapranov-Zelevinsky construction of the associahedron as a "secondary polytope" is given by G. Ziegler, *Lectures on Polytopes*, Springer-Verlag, New York, 1995. For a more recent survey paper on the associahedron, see C. Ceballos, F. Santos, and G. Ziegler, *Combinatorica*, to appear; `arXiv:1109.5544`.

(c) See J.-L. Loday, *Archiv der Math.* **83** (2004), 267–278.

(d) Write

$$F(x,y) = x + \sum_{n \geq 2} \sum_{i=1}^{n-1} W_{i-1}(n+1) x^n y^i$$

$$= x + x^2 y + x^3(y + 2y^2) + x^4(y + 5y^2 + 5y^3) + \cdots.$$

By removing a fixed exterior edge from a dissected polygon and considering the edge-disjoint union of polygons thus formed, we get the functional equation

$$F = x + y \frac{F^2}{1 - F}.$$

(Compare [65, (6.15)].) Hence by [65, Exercise 5.59] we have

$$F = \sum_{m \geq 1} \frac{1}{m} [t^{m-1}] \left(x + y \frac{t^2}{1-t} \right)^m$$

$$= \sum_{m \geq 1} \frac{1}{m} [t^{m-1}] \sum_{n=0}^{m} \binom{m}{n} x^n \left(y \frac{t^2}{1-t} \right)^{m-n}.$$

From here it is a simple matter to obtain

$$F = x + \sum_{n \geq 2} \sum_{i=1}^{n-1} \frac{1}{n+i} \binom{n+i}{i} \binom{n-2}{i-1} x^n y^i,$$

whence

$$W_i(n) = \frac{1}{n+i} \binom{n+i}{i+1} \binom{n-3}{i}. \tag{5.3}$$

This formula goes back to T. P. Kirkman, *Phil. Trans. Royal Soc. London* **147** (1857), 217–272; E. Prouhet, *Nouvelles Annales Math.* **5** (1866), 384; and A. Cayley, *Proc. London Math. Soc. (1)* **22** (1890–1891), 237–262, who gave the first complete proof. For Cayley's proof see also [24, Ch. 7.3]. For a completely different proof, see [65, Exer. 7.17]. Another proof appears in D. Beckwith, *Amer. Math. Monthly* **105** (1998), 256–257.

(e) We have $h_i = \frac{1}{n-1} \binom{n-3}{i} \binom{n-1}{i+1}$, a result of C. W. Lee, in *DIMACS Series in Discrete Mathematics and Theoretical Computer Science*, vol. 4, 1991, pp. 443–456.

A43. (a,d,e) See J. Fürlinger and J. Hofbauer, *J. Combinatorial Theory (A)* **40** (1985), 248–264, and the references given there. In particular, (a) is due to L. Carlitz, *Fibonacci Quart.* **10** (1972), 531–549, while (d) is due to MacMahon, *Phil. Trans.* **211** (1912), 75–110 (p. 92), reprinted in *Collected Papers*, G. Andrews, ed., M.I.T. Press, Cambridge, Massachusetts, 1978, pp. 1328–1363, and [48, §465]. For (c) see also G. E. Andrews, *J. Stat. Plan. Inf.* **34** (1993), 19–22 (Corollary 1). The continued fraction (4.14) is of a type considered by Ramanujan. It is easy to show (see for instance G. E. Andrews [4, §7.1]) that

$$F(x) = \frac{\displaystyle\sum_{n \geq 0} (-1)^n q^{n^2} \frac{x^n}{(1-q)\cdots(1-q^n)}}{\displaystyle\sum_{n \geq 0} (-1)^n q^{n(n-1)} \frac{x^n}{(1-q)\cdots(1-q^n)}}.$$

(b) See the last paragraph of C. Bessenrodt and R. Stanley, *J. Alg. Comb.*, to appear; arXiv:1311.6123. The determinants had earlier been found by J. Cigler, *Sitzungsber. OAW* **208** (1999), 3–20, and preprint, arXiv:math.CO/0507225.

(c) See R. Bacher and C. Reutenauer, in *Contemporary Math.* **592** (2013), pp. 1–18 (Theorem 1). The authors also give a Fuss-Catalan generalization of this result.

(f,g) See R. Stanley, *Ann. New York Acad. Sci.* **574** (1989), 500–535
(Example 4 on p. 523). The unimodality result is closely related
to item A8(c).

A44. (a) The bounce statistic and its connection with the algebra B are due to
J. Haglund. See his monograph *The q,t-Catalan numbers and the
space of diagonal harmonics*, University Lecture Series, vol. 41,
American Mathematical Society, Providence, Rhode Island, 2008,
for an exposition of this subject. We have taken the definition
of bounce almost verbatim from this monograph. The polynomial
$F_n(q,t)$ is known as a (q,t)-*Catalan number.*

A45. These results are due to V. Welker, *J. Combinatorial Theory (B)* **63**
(1995), 222–244 (§4).

A46. (a) There are $\binom{n}{k-1}\binom{n-1}{k-1}$ pairs of compositions $A : a_1 + \cdots + a_k = n+1$
and $B : b_1 + \cdots + b_k = n$ of $n+1$ and n into k parts. Construct
from these compositions a circular sequence $w = w(A,B)$ consisting
of a_1 1's, then b_1 -1's, then a_2 1's, then b_2 -1's, etc. Because n
and $n+1$ are relatively prime, this circular sequence w could have
arisen from exactly k pairs (A_i, B_i) of compositions of $n+1$ and n
into k parts, viz., $A_i : a_i + a_{i+1} + \cdots + a_k + a_1 + \cdots + a_{i-1} = n+1$
and $B_i : b_i + b_{i+1} + \cdots + b_k + b_1 + \cdots + b_{i-1} = n$, $1 \leq i \leq k$. By
Lemma 1.6.1 there is exactly one way to break w into a linear
sequence $\overline{w}$ such that $\overline{w}$ begins with a 1, and when this initial 1 is
removed every partial sum is nonnegative. Clearly there are exactly
k 1's in $\overline{w}$ followed by a -1 (with or without the initial 1 of $\overline{w}$
removed). This sets up a bijection between the set of all "circular
equivalence classes" $\{(A_1, B_1), \ldots, (A_k, B_k)\}$ and X_{nk}. Hence

$$X_{nk} = \frac{1}{k}\binom{n}{k-1}\binom{n-1}{k-1} = \frac{1}{n}\binom{n}{k-1}\binom{n}{k}.$$

(b) For $k \geq 1$, let $X_k = X_{1k} \cup X_{2k} \cup \cdots$. Every $w \in X_k$ can be written
uniquely in one of the forms (i) $1\,u\,-1\,v$, where $u \in X_j$ and $v \in X_{k-j}$
for some $1 \leq j \leq k-1$, (ii) $1-1\,u$, where $u \in X_{k-1}$, (iii) $1\,u\,-1$, where
$u \in X_k$, and (iv) $1-1$ (when $k = 1$). Regarding X_k as a language as in
[65, Example 6.6.6], and replacing for notational comprehensibility
1 by α and -1 by β, conditions (i)–(iv) are equivalent to the equation

$$X_k = \sum_{j=1}^{k-1} \alpha X_j \beta X_{k-j} + \alpha\beta X_{k-1} + \alpha X_k \beta + \delta_{1k}\alpha\beta.$$

Thus if $y_k = \sum_{n \geq 1} N(n,k)x^n$, it follows that (setting $y_0 = 0$)

$$y_k = x \sum_{j=0}^{k} y_j y_{k-j} + xy_{k-1} + xy_k + \delta_{1k}x.$$

Since $F(x,t) = \sum_{k \geq 1} y_k t^k$, we get (4.16).

Narayana numbers were introduced by P. A. MacMahon [48, §495] (in fact, a q-analogue). They were rediscovered by T. V. Narayana, *C. R. Acad. Sci. Paris* **240** (1955), 1188–1189, and considered further by him in *Sankhya* **21** (1959), 91–98; and *Lattice Path Combinatorics with Statistical Applications*, Mathematical Expositions no. 23, University of Toronto Press, Toronto, 1979 (Section V.2). Further references include G. Kreweras and P. Moszkowski, *J. Statist. Plann. Inference* **14** (1986), 63–67; G. Kreweras and Y. Poupard, *European J. Combinatorics* **7** (1986), 141–149; R. A. Sulanke, *Bull. Inst. Combin. Anal.* **7** (1993), 60–66; and R. A. Sulanke, *J. Statist. Plann. Inference* **34** (1993), 291–303.

A47. Statement (b) was conjectured by N. Wormald at a meeting in Oberwolfach in March, 2014. Statements (a) and (b) were proved independently by I. Gessel, G. Schaeffer, and R. Stanley by a combinatorial argument involving exponential generating functions. Gessel then went on to prove (c) and connect these results with continued fractions in a preprint, "Two solutions to Nick Wormald's problem," dated April 2, 2014.

A48. Equivalent to [64, Exercise 1.155(c)]. We thus have that C_n is the unique function $f(n)$ such that $f(0) = 1$, and if $g(n) = \Delta^n f(0)$ (where Δ is the difference operator) then $\Delta^{2n} g(0) = f(n)$ and $\Delta^{2n+1} g(0) = 0$. See also the nice survey by R. Donaghey and L. W. Shapiro, *J. Combinatorial Theory (A)* **23** (1977), 291–301.

A49. All these results not discussed below appear in Donaghey and Shapiro, *ibid.* Donaghey and Shapiro give several additional interpretations of Motzkin numbers and state that they have found a total of about forty interpretations. The name "Motzkin number" arose from the paper T. Motzkin, *Bull. Amer. Math. Soc.* **54** (1948), 352–360.

(f) See M. S. Jiang, in *Combinatorics and Graph Theory* (Hefei, 1992), World Scientific Publishing, River Edge, New Jersey, 1993, pp. 31–39.

(k) See A. Kuznetsov, I. Pak, and A. Postnikov, *J. Combinatorial Theory (A)* **76** (1996), 145–147.

(l) See M. Aigner, *Europ. J. Combinatorics* **19** (1998), 663–675. Aigner calls the partitions of part (l) *strongly noncrossing*.

(m) See M. Klazar, *Europ. J. Combinatorics* **17** (1996), 53–68 (pp. 55–56) (and compare [64, Exercise 1.108(a)]). Klazar's paper contains a number of further enumeration problems related to the present one that lead to algebraic generating functions; see items 162,163 and A50(r) for three of them.

(n) See O. Guibert, E. Pergola, and R. Pinzani, *Ann. Combinatorics* **5** (2001), 153–174.

(o) Let f_n be the number of such trees on n vertices. Then

$$f_{n+1} = \sum_{k=1}^{n} f_k f_{n-k}, \quad n \geq 1.$$

Setting $y = \sum_{n \geq 0} f_n x^n$ we obtain

$$\frac{y - 1 - x}{x} = y^2 - y.$$

It follows that

$$y = \frac{1 + x - \sqrt{1 - 2x - 3x^2}}{2x}.$$

Comparing with the definition of M_n in Problem A48 shows that $f_n = M_{n-1}$, $n \geq 1$. It shouldn't be difficult to give a bijective proof.

(p) It is known (e.g., Lemma 2.1 of S. Dulucq and R. Simion, *J. Alg. Comb.* **8** (1998), 169–191) that a permutation $w \in \mathfrak{S}_n$ has genus 0 if and only if each cycle of w is increasing (when the smallest element of the cycle is written first) and the sets of elements of the cycles (i.e., the orbits of w) form a noncrossing partition of $[n]$. It follows that both w and w^{-1} have genus 0 if and only if w is an involution whose cycle elements form a noncrossing (partial) matching. The proof now follows from part (a). This item is due to E. Deutsch, private communication dated August 26, 2009.

A50. (a) See [65, §6.2].

(b,e,f,j,k) These follow from part (a) using the bijections of the proof of Theorem 1.5.1.

(c) See D. Gouyou-Beauchamps and D. Vauquelin, *RAIRO Inform. Théor. Appl.* **22** (1988), 361–388. This paper gives some other tree representations of Schröder numbers, as well as connections with Motzkin numbers and numerous references.

(d) An easy consequence of the paper of Shapiro and Stephens cited below.

(g) Rotate the paths of (l) 45° clockwise, reflect about the x-axis, and double coordinates. Both the paths of (l) and the present item are known as (large) *Schröder paths*.

(h) Let L be a path counted by (g) with no level steps on the x-axis. For every up step s beginning at the x-axis, let t be the first down step after s ending at the x-axis. Remove the steps s and t, lower the portion of L between s and t by one unit, and append a step $(2, 0)$ at the end of this lowered portion of L. We obtain a bijection between paths counted by the present problem and those paths in (t) with at least one level step on the x-axis. Hence the total number paths counted by the present problem is $\frac{1}{2} r_n = s_n$. The paths of this item are known as *small Schröder paths*.

(i) Due to R. A. Sulanke, *J. Difference Equations Appl.* **5** (1999), 155–176. The objects counted by this problem are called *zebras*. See also E. Pergola and R. A. Sulanke, *J. Integer Sequences* (electronic) **1** (1998), Article 98.1.7.

(l,m) See L. W. Shapiro and A. B. Stephens, *SIAM J. Discrete Math.* **4** (1991), 275–280. For part (l), see also [65, Exercise 6.17(b)].

(n) L. W. Shapiro and S. Getu (unpublished) conjectured that the set $\mathfrak{S}_n(2413, 3142)$ and the set counted by (m) are identical (identifying a permutation matrix with the corresponding permutation). It was proved by J. West, *Discrete Math.* **146** (1995), 247–262, that $\#\mathfrak{S}_n(2413, 3142) = r_{n-1}$. Since it is easy to see that permutations counted by (m) are 2413-avoiding and 3142-avoiding, the conjecture of Shapiro and Getu follows from the fact that both sets have cardinality r_{n-1}. Presumably there is some direct proof that the set counted by (m) is identical to $\mathfrak{S}_n(2413, 3142)$.

West also showed in Theorem 5.2 of the above-mentioned paper that the sets $\mathfrak{S}_n(1342, 1324)$ and (m) are identical. The enumeration of $\mathfrak{S}_n(1342, 1432)$ was accomplished by S. Gire, PhD thesis, Université Bordeaux, 1991. The remaining seven cases were enumerated by D. Kremer, *Discrete Math.* **218** (2000), 121–130, and **270** (2003), 333–334. Kremer also gives proofs of the three previously known cases. She proves all ten cases using the method of "generating trees" introduced by F.R.K. Chung, R. L. Graham, V. E. Hoggatt, Jr., and M. Kleiman, *J. Combinatorial Theory (A)* **24**

(1978), 382–394, and further developed by J. West, *Discrete Math.* **146** (1995), 247–262, and **157** (1996), 363–374. It has been verified by computer that there are no other pairs $(u, v) \in \mathfrak{S}_4 \times \mathfrak{S}_4$ (up to equivalence) for which $\#\mathfrak{S}_n(u, v) = r_{n-1}$ for all n.

(o) This is a result of Knuth [33, Exercise 2.2.1.10–2.2.1.11, pp. 239 and 533–534]; these permutations are now called *deque-sortable*. A combinatorial proof appears in D. G. Rogers and L. W. Shapiro, in *Lecture Notes in Math.*, no. 884, Springer-Verlag, Berlin, 1981, pp. 293–303. Some additional combinatorial interpretations of Schröder numbers and many additional references appear in this reference. For q-analogues of Schröder numbers, see J. Bonin, L. W. Shapiro, and R. Simion, *J. Stat. Planning and Inference* **34** (1993), 35–55.

(p) See K. Mészáros, *Trans. Amer. Math. Soc.* **363** (2011), 4359–4382 (§7).

(q) See D. G. Rogers and L. W. Shapiro, *Lecture Notes in Mathematics*, no. 686, Springer-Verlag, Berlin, 1978, pp. 267–276 (§5) for simple bijections with (a) and other "Schröder structures."

(r) This result is due to R. C. Mullin and R. G. Stanton, *Pacific J. Math.* **40** (1972), 167–172 (§3), using the language of "Davenport-Schinzel sequences." It is also given by M. Klazar, *Europ. J. Combinatorics* **17** (1996), 53–68 (p. 55).

(s) Remove a "root edge" from the polygon of (j) and "straighten out" to obtain a noncrossing graph of the type being counted.

(t,u) These results (which despite their similarity are not trivially equivalent) appear in D. G. Rogers, *Lecture Notes in Mathematics*, no. 622, Springer-Verlag, Berlin, 1977, pp. 175–196 (equations (38) and (39)), and are further developed in D. G. Rogers, *Quart. J. Math. (Oxford) (2)* **31** (1980), 491–506. In particular, a bijective proof that (t) and (u) are equinumerous appears in Section 3 of this latter reference. It is also easy to see that (s) and (u) are virtually identical problems. A further reference is D. G. Rogers and L. W. Shapiro, *Lecture Notes in Mathematics*, no. 686, Springer-Verlag, Berlin, 1978, pp. 267–276.

(v) See M. Ciucu, *J. Algebraic Combinatorics* **5** (1996), 87–103, (Theorem 4.1).

(w) Walk from left to right along a path of (h). Insert the least unused positive integer in the first row (as far to the left as possible) if an up step is encountered, in the second row if a down step is encountered, and in both the top and bottom rows if a level step is encountered, thus setting up a bijection with the matrices being counted. See O. Pechenik, *J. Combinatorial Theory Ser. A* **125** (2014), 357–378, for some additional aspects.

(x) Let $f(n)$ be the number of guillotine rectangulations being counted. Exactly $f(n)/2$ guillotine rectangulations contain a vertical line segment cutting R into two rectangles. By considering the first point p_i left to right through which such a vertical line segment passes, we obtain the recurrence

$$\frac{1}{2}f(n) = f(n-1) + \sum_{k=2}^{n} \left(\frac{1}{2}f(k-1) \right) f(n-k).$$

It is straightforward to obtain from this recurrence that $f(n) = r_n$. This result is due to E. Ackerman, G. Barequet, and R. Y. Pinter, *J. Combinatorial Theory, Ser. A* **113** (2006), 1072–1091 (Theorem 2).

(y) It was shown by F. Wei, `arXiv:1009.5740`, that the 2413 and 3142-avoiding permutations (also called *separable permutations*) included in (n) have the stated property. It is conjectured that these are the only such permutations.

(z) See R. Johansson and S. Linusson, *Ann. Combinatorics* **11** (2007), 471–480 (§3.4), where a bijection is given with (l).

(aa) In fact, these are just the 2413 and 3142-avoiding (or separable) permutations of (y). See E. Ghys, *Amer. Math. Monthly* **120** (2013), 232–242.

A51. Note that this problem is the "opposite" of Problem A50(m), i.e., here we are counting the permutation matrices P for which not even a single new 1 can be added (using the rules of Problem A50(m)). The present problem was solved by Shapiro and Stephens in Section 3 of the paper cited in the solution to Problem A50(m). For a less elegant form of the answer and further references, see M. Abramson and W.O.J. Moser, *Ann. Math. Stat.* **38** (1967), 1245–1254.

A52. This result was conjectured by J. West, PhD thesis, MIT, 1990 (Conjecture 4.2.19), and first proved by D. Zeilberger, *Discrete Math.* **102** (1992), 85–93. For further proofs and related results, see M. Bóna,

2-stack sortable permutations with a given number of runs, preprint dated May 13, 1997; M. Bousquet-Mélou, *Electron. J. Combinatorics* **5** (1998), R21, 12pp.; S. Dulucq, S. Gire, and J. West, *Discrete Math.* **153** (1996), 85–103; I. P. Goulden and J. West, *J. Combinatorial Theory (A)* **75** (1996), 220–242; J. West, *Theoret. Comput. Sci.* **117** (1993), 303–313; and M. Bousquet-Mélou, *Elec. J. Combinatorics* **5(1)** (1998), R21, 12pp.

A53. This result is due to P. H. Edelman and V. Reiner, *Graphs and Combinatorics* **13** (1997), 231–243 (Theorem 1).

A54. These remarkable results are due to G. Viennot and D. Gouyou-Beauchamps, *Advances in Appl. Math.* **9** (1988), 334–357. The subsets being enumerated are called *directed animals*. For a survey of related work, see G. Viennot, *Astérisque* **121–122** (1985), 225–246. See also the two other papers cited in the solution to item 194, as well as the paper M. Bousquet-Mélou, *Discrete Math.* **180** (1998), 73–106. Let us also mention that Viennot and Gouyou-Beauchamps show that $f(n)$ is the number of sequences of length $n-1$ over the alphabet $\{-1,0,1\}$ with nonnegative partial sums. Moreover, M. Klazar, *Europ. J. Combinatorics* **17** (1996), 53–68 (p. 64), shows that $f(n)$ is the number of partitions of $[n+1]$ such that no block contains two consecutive integers, and such that if $a < b < c < d$, a and d belong to the same block B_1, and b and c belong to the same block B_2, then $B_1 = B_2$. For a variant, see item 194.

A55. (a) There is a third condition equivalent to (i) and (ii) of (a) that motivated this work. Every permutation $w \in \mathfrak{S}_n$ indexes a closed Schubert cell $\overline{\Omega}_w$ in the complete flag variety $\mathrm{GL}(n,\mathbb{C})/B$. Then w is smooth if and only if the variety $\overline{\Omega}_w$ is smooth. The equivalence of this result to (i) and (ii) is implicit in K. M. Ryan, *Math. Ann.* **276** (1987), 205–244, and is based on earlier work of Lakshmibai, Seshadri, and Deodhar. An explicit statement that the smoothness of $\overline{\Omega}_w$ is equivalent to (ii) appears in V. Lakshmibai and B. Sandhya, *Proc. Indian Acad. Sci. (Math. Sci.)* **100** (1990), 45–52. See also S. C. Billey and V. Lakshmibai, *Singular Loci of Schubert Varieties*, Birkhäuser Boston, Boston, Massachusetts, 2000.

(b) This generating function is due to M. Haiman (unpublished).

NOTE. It was shown by M. Bóna, *Electron. J. Combinatorics* **5** (1998), R31, that there are four other inequivalent (in the sense of Problem A50(n)) pairs $(u,v) \in \mathfrak{S}_4 \times \mathfrak{S}_4$ such that the number of permutations in $\mathfrak{S}_n$ that avoid them is equal to $f(n)$, viz., $(1324, 2413)$, $(1342, 2314)$, $(1342, 2431)$, and $(1342, 3241)$. (The case $(1342, 2431)$ is implicit in Z. Stankova, *Discrete Math.* **132** (1994), 291–316.)

A56. Setting $b_k = 0$ for $k \notin \mathbb{P}$, it is easy to obtain the recurrence

$$b_n = \sum_{\substack{i+j=n \\ i<j}} b_i b_j + \left(\binom{b_{n/2}}{2} \right), \quad n \geq 2,$$

from which (4.17) follows readily. This problem was considered by J.H.M. Wedderburn, *Ann. Math.* **24** (1922), 121–140, and I.M.H. Etherington, *Math. Gaz.* **21** (1937), 36–39, and is known as the *Wedderburn-Etherington commutative bracketing problem*. For further information and references, see L. Comtet, *Advanced Combinatorics*, Reidel, Boston, 1974 (pp. 54–55); and H. W. Becker, *Amer. Math. Monthly* **56** (1949), 697–699.

A57. This result is due to V. Liskovets and R. Pöschel, *Discrete Math.* **214** (2000), 173–191. For the case $p = 2$, see I. Kovács, *Sém. Lotharingien de Combinatoire* (electronic) **51** (2005), Article B51h.

A58. See F. Ardila, *J. Combinatorial Theory, Ser. A* **104** (2003), 49–62; and J. E. Bonin, A. de Mier, and M. Noy, *J. Combinatorial Theory, Ser. A* **104** (2003), 63–94.

A59. (a) *Answer:* $f(x) = \sqrt{4 - x^2}/2\pi$ for $-2 \leq x \leq 2$, and $f(x) = 0$ for $|x| \geq 2$. This result is the basis of Wigner's famous "semicircle law" for the distribution of eigenvalues of certain classes of random real symmetric matrices (*Ann. Math.* **62** (1955), 548–569, and **67** (1958), 325–327). Wigner did not rigorously prove the uniqueness of $f(x)$, but this uniqueness is actually a consequence of earlier work of F. Hausdorff, *Math. Z.* **16** (1923), 220–248.

(b) *Answer:* $f(x) = \frac{1}{2\pi}\sqrt{\frac{4-x}{x}}$ for $0 \leq x \leq 4$, and $f(x) = 0$ otherwise, an easy consequence of (a).

A60. It is not difficult to see that the term indexed by the composition $a_0 + a_1 + \cdots + a_k$ is the number of sequences counted by item 95 such that i appears a_i times. This result is due to D. Zare, `mathoverflow.net/questions/131585`.

A61. (a) Write $C_n = \frac{1}{2n+1}\binom{2n+1}{n}$. Since $2n + 1$ is odd, it follows from Kummer's theorem that the exponent of the largest power of 2 dividing C_n is equal to the number of carries in adding n and $n + 1$ in base 2. It is not difficult to see that this number is $b(n + 1) - 1$. This result essentially dates back to Kummer.

(b) See E. Deutsch and B. Sagan, *J. Number Theory* **117** (2006), 191–215.

(c) See M. Kauers, C. Krattenthaler, and T. W. Müller, *Electr. J. Comb.* **18(2)** (2012), #P37 (equation (1.2)). In Section 3 of this paper it is explained why this result gives an efficient determination of C_n (mod 64). Similar formulas exist for C_n (mod 2^m) for any $m \geq 1$.

A62. These results are due to C. B. Pomerance, On numbers related to Catalan numbers, preprint, 2013.

A63. *Idea of proof.* As $n \to \infty$, an integer $1 \leq i \leq n$ will have probability $1/2$ of having an odd number of 1's in its binary expansion. Hence by Problem A61, as $n \to \infty$ a Catalan number C_i with $1 \leq i \leq n$ will be of the form $4^a(2c+1)$ with probability $1/2$. When this happens, the residue of $2c + 1$ modulo 8 will be equidistributed among the four odd residue classes as $n \to \infty$. Thus the probability that C_i is not a sum of three squares is $\frac{1}{2} \cdot \frac{1}{4}$ as $n \to \infty$.

A64. Straightforward. More strongly, Stirling's asymptotic series for $n!$ leads to the asymptotic series

$$C_n = \frac{4^n}{\sqrt{\pi}\, n^{3/2}} \left(1 - \frac{9}{8n} + \frac{145}{128n^2} + \frac{5}{1024n^3} - \frac{21}{32768n^4} + \cdots \right).$$

A65. (a) Once Equation (4.18) is known or conjectured, it is routine to verify it using a suitable binomial coefficient identity. It can also easily be deduced from known series expansions of $\left(\sin^{-1}x\right)^2$ or $(\sin^{-1}x)/\sqrt{1-x^2}$. See for instance D. H. Lehmer, *Amer. Math. Monthly* **92** (1985), 449–457, for some additional information and references. A direct proof is given by T. Koshy and Z. Gao, in *Martin Gardner in the Twenty-First Century* (M. Henle and B. Hopkins, eds.), Mathematical Association of America, Washington, DC, 2012, pp. 119–124. For some analytic aspects see F. J. Dyson, N. E. Frankel, and M. L. Glasser, *Amer. Math. Monthly* **120** (2013), 116–130.

(b) Set $x = 1$ in equation (4.18) and use $\sin^{-1} \frac{1}{2} = \frac{\pi}{6}$.

(c) Let $F(x)$ denote the right-hand side of (4.18). Compute $4F(1) - 3F'(1)$.

A66. The series $\sum C_n 4^{-n}$ converges by Problem A64. Now apply Abel's theorem to $C(x/4)$, where $C(x)$ is given by equation (1.3).

A67. (a) Perhaps the most straightforward proof is obtained by differentiating the claimed identity, yielding

$$\frac{C'(x)}{C(x)} = \sum_{n \geq 1} \binom{2n-1}{n-1} x^{n-1} = \frac{1}{2x} \left(-1 + \frac{1}{\sqrt{1-4x}} \right). \tag{5.4}$$

Equation (5.4) can be easily verified using the formula (1.3) for $C(x)$. The proof then follows by checking that the coefficient of x on both sides of equation (4.19) is the same. For a combinatorial proof, see L. W. Shapiro, *Congr. Numerantium* **199** (2009), 217–222. For a q-analogue see Problem A45.

(b) Let $f(n)$ denote the number of ways to choose a forest F of plane trees on the vertices $1, 2 \ldots, n$, and then arrange the connected components (i.e., the plane trees) of F into a cycle. Call this structure a *cyclic plane forest*. It can be shown combinatorially that $f(n) = (n-1)!\binom{2n-1}{n} = (2n-1)(2n-2) \cdots (n+1)$. Now choose a partition $\pi = \{B_1, \ldots, B_k\}$ of $1, 2, \ldots, n$ and place a cyclic plane forest on each block B_i. We thus have a plane forest on $[n]$ whose components are arranged into a disjoint union of cycles. A disjoint union of cycles corresponds to a permutation (linear ordering) $T_1, \ldots, T_k$ of the components. Adjoin a new vertex v and connect it to the root of each tree T_i. Regarding v as the root of the resulting tree, we now have a plane tree T on $n+1$ vertices whose nonroot vertices are labelled $1, 2, \ldots, n$. There are C_n ways to choose T (item 6) and $n!$ ways to label the nonroot vertices. Hence

$$n! C_n = \sum_{\pi = \{B_1, \ldots, B_k\} \in \Pi_n} f(\#B_1) \cdots f(\#B_k),$$

and the proof follows from the exponential formula. This proof was obtained in collaboration with W. Chen.

A68. See items A001517 and A080893 of *The On-line Encyclopedia of Integer Sequences*, oeis.org.

Appendix A

In the beginning...

The next four pages show a very early version of this monograph, from the 1970s.

CATALAN NUMBERS

$$C_m = \frac{1}{m+1}\binom{2m}{m} \qquad C_1 = 1,\ C_2 = 2$$
$$C_3 = 5,\ C_4 = 14$$

1. $e(\underline{2} \times \underline{m})$

2. no. of lattice paths in an $(m+1) \times (m+1)$ grid not going below diagonal

3. no. of order ideals of $\delta(\underline{m-1})^{-\{\phi\}}$ (or $[0, \lambda]$ in $J(\underline{N}^2)$, where $\lambda = (m-1, m-2, \ldots, 1)$)

4. no. of ways of parenthesizing $m+1$ factors

5. no. of ways of dividing an $m+2$-gon into triangles by non-intersecting diagonals

6. no. of non-isomorphic ordered sets of card m with no sub-ordered sets $\begin{smallmatrix}\bullet&\bullet\\\bullet&\bullet\end{smallmatrix}$ or $\begin{smallmatrix}\bullet\\\bullet\\\bullet\end{smallmatrix}$

7. no. of permutations of $1, 2, \ldots, m$ with longest increasing subsequence of length ≤ 2

8. no. of two-sided ideals in the algebra of $(m-1) \times (m-1)$ upper triangular matrices over $\mathbb{R}$

9. no. of sequences $1 \leq a_1 \leq \cdots \leq a_m$ with $a_i \leq i$

10. no. of sequences $\epsilon_1, \epsilon_2, \ldots, \epsilon_{2m}$ of ± 1's with every partial sum $s_k \geq 0$ and $s_{2m} = 0$ (ballot problem)

11. no. of size sequences of principal ideals of posets

12. Berlekamp determinant with 1~1 boundary

13. no. of plane binary trees with n vertices
(order ideal interpretation)

14. no. of plane planted trees with $n+1$ vertices

15. no. of partitions of $\{1, 2, \ldots, n\}$ such that if ~~a~b and c~d~~ $a < b < c < d$, then we never have $a \sim c$ and $b \sim d$
unless $a \sim b \sim c \sim d$

16. ~~$\mu(0,1)$ for the~~ $(-1)^{n-1} \mu(0,1)$ for the ordered set of partitions of $\{1, 2, \ldots, n+1\}$ satisfying (15)

17. no. of ways $2n$ points on the circumference of a circle can be joined in pairs by n non-intersecting chords

18. no. of planted (root has degree 1) trivalent plane trees on $2n+2$ vertices

19. no. of n-tuples $a_1, \ldots, a_n$, $a_i \in \mathbb{P}$, such that in the sequence $1 \, a_1 \, a_2 \cdots a_n \, 1$, each a_i divides the sum of its two neighbors

20. no. of permutations $a_1, \ldots, a_n$ of $[n]$ with no subsequence a_i, a_j, a_k $(i < j < k)$ satisfying $a_j < a_k < a_i$

21. no. of permutations $a_1, \ldots, a_{2n}$ of the multiset $\{1^2, 2^2, \ldots, n^2\}$ such that: (i) first occurrences of $1, \ldots, n$ appear in increasing order, (ii) no subsequence of the form $\alpha\beta\alpha\beta$. (The second occurrences of $1, \ldots, n$ form a permutation as in 20.)

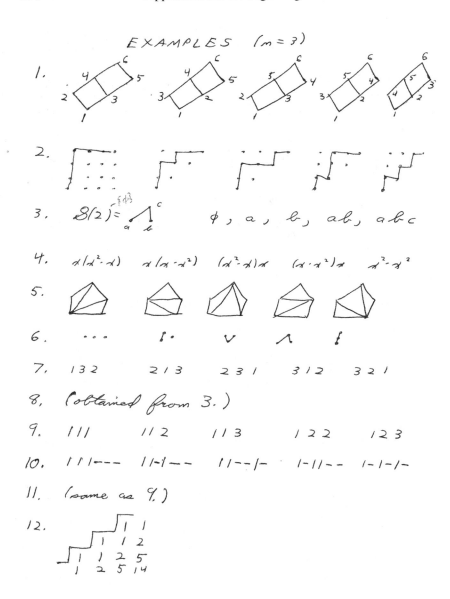

EXAMPLES (m = 3)

1.

2.

3. $8(2) = $ ϕ, a, b, ab, abc

4. $a(a^2 \cdot a)$ $a(a \cdot a^2)$ $(a^2 \cdot a)a$ $(a \cdot a^2)a$ $a^2 \cdot a^2$

5.

6. . . .   .   ∨   ∧

7. 1 3 2   2 1 3   2 3 1   3 1 2   3 2 1

8. (obtained from 3.)

9. 1 1 1   1 1 2   1 1 3   1 2 2   1 2 3

10. 1 1 1 - - -   1 1 - 1 - -   1 1 - - 1 -   1 - 1 1 - -   1 - 1 - 1 -

11. (same as 9.)

12.

13.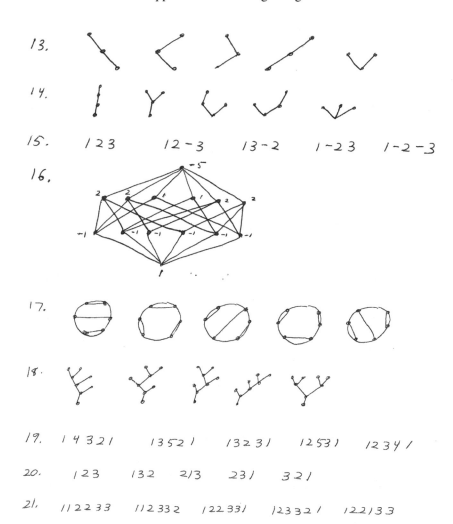

14.

15.   1 2 3   1 2 - 3   1 3 - 2   1 - 2 3   1 - 2 - 3

16.

17.

18.

19.   1 4 3 2 1   1 3 5 2 1   1 3 2 3 1   1 2 5 3 1   1 2 3 4 1

20.   1 2 3   1 3 2   2 / 3   2 3 1   3 2 1

21.   1 1 2 2 3 3   1 1 2 3 3 2   1 2 2 3 3 1   1 2 3 3 2 1   1 2 2 1 3 3

Appendix B
History of Catalan Numbers (by Igor Pak)

Introduction

In the modern mathematical literature, Catalan numbers are wonderfully ubiquitous. Although they appear in a variety of disguises, we are so used to having them around, it is perhaps hard to imagine a time when they were either unknown, or known but obscure and underappreciated. It may then come as a surprise that Catalan numbers have a rich history of multiple rediscoveries until relatively recently. Here we review more than 200 years of history, from their first discovery to modern times.

We break the history into short intervals of mathematical activity, each covered in a different section. We spend most of our effort on the early history but do bring it to modern times. We should warn the reader that although this work is in the History of Mathematics, the author is not a mathematical historian. Rather, this work is more of a historical survey with some added speculations based on our extensive reading of the even more extensive literature. Due to the space limitations, this survey is very much incomplete, as we tend to emphasize first discoveries and papers of influence rather than describe subsequent developments.

This paper in part is based on our earlier investigation reported in [54]. Many primary sources are assembled on the *Catalan Numbers website* [55], including scans of the original works and their English translations.

B.1. Ming Antu

The Mongolian astronomer, mathematician, and topographic scientist Minggatu (full name Sharabiin Myangat) (c. 1692–c. 1763), worked at the Qing court in China. Ming's Chinese name is Ming'antu and courtesy name is Jing An. In the 1730s, he wrote a book *Quick Methods for Accurate Values of Circle Segments*, which included a number of trigonometric identities and power series, some involving Catalan numbers:

$$\sin(2\alpha) = 2\sin\alpha - \sum_{n=1}^{\infty} \frac{C_{n-1}}{4^{n-1}} \sin^{2n+1}\alpha = 2\sin\alpha - \sin^3\alpha - \frac{1}{4}\sin^5\alpha - \frac{1}{8}\sin^7\alpha - \cdots$$

He also obtained the recurrence formula

$$C_1 = 1, \quad C_2 = 2, \quad C_{n+1} = \sum_{k \geq 0} (-1)^k \binom{n+1-k}{k+1} C_{n-k}.$$

He appears to have no inkling of a combinatorial interpretation of Catalan numbers.

Ming Antu's book was published only in 1839, and the connection to Catalan numbers was observed by Luo Jianjin in 1988. For further information, see J. J. Luo [47] and P. J. Larcombe [40].

B.2. Euler and Goldbach

In 1751, Leonhard Euler (1707–1783) introduced and found a closed formula for what we now call the *Catalan numbers*. The proof of this result had eluded him, until he was assisted by Christian Goldbach (1690–1764), and more substantially by Johann Segner. By 1759, a complete proof was obtained. This and the next section tell the story of how this happened.

On September 4, 1751, Euler wrote a letter to Goldbach which among other things included his discovery of Catalan numbers.[1] Euler was in Berlin (Prussia) at that time, while his friend and former mentor Goldbach was in St. Petersburg (Imperial Russia). They first met when Euler arrived to St. Petersburg back in 1727 as a young man. This meeting started a lifelong friendship, with 196 letters written between them [71].

Euler defines Catalan numbers C_n as the number of triangulations of an $(n+2)$-gon, and gives the values of C_n for $n \leq 8$ (evidently computed by hand). All these values, including $C_8 = 1430$, are correct. Euler then observes that successive ratios have a pattern and guesses the following formula for Catalan numbers:

$$C_{n-2} = \frac{2 \cdot 6 \cdot 10 \cdots (4n-10)}{2 \cdot 3 \cdot 4 \cdots (n-1)}. \tag{B.1}$$

---

[1] The letters between Euler and Goldbach were published in 1845 by Paul Heinrich von Fuss (or Pavel Nikolaevich Fuss), the son of Nicolaus (or Nikolai) Fuss. The letters were partly excised, but the originals also survived. Note that the first letter shows labeled quadrilateral and pentagon figures missing in the published version; see [55].

For example, $C_3 = (2 \cdot 6 \cdot 10)/(2 \cdot 3 \cdot 4) = 5$. He concludes with the formula for the Catalan number generating function:

$$A(x) = 1 + 2x + 5x^2 + 14x^3 + 42x^4 + 132x^5 + \ldots = \frac{1 - 2x - \sqrt{1 - 4x}}{2x^2}. \quad (B.2)$$

In his reply to Euler dated October 16, 1751, Goldbach notes that the generating function $A(x)$ satisfies the quadratic equation

$$1 + xA(x) = A(x)^{\frac{1}{2}}. \quad (B.3)$$

He then suggests that this equation can be used to derive Catalan numbers via an infinite family of equations on its coefficients.

Euler writes back to Goldbach on December 4, 1751. He explains how one can obtain the product formula (B.1) from the binomial formula:

$$\sqrt{1 - 4x} = 1 - \frac{1}{2}4x - \frac{1 \cdot 1}{2 \cdot 4}4^2 x^2 - \frac{1 \cdot 1 \cdot 3}{2 \cdot 4 \cdot 6}4^3 x^3 - \frac{1 \cdot 1 \cdot 3 \cdot 5}{2 \cdot 4 \cdot 6 \cdot 8}4^4 x^4 - \ldots \quad (B.4)$$

From the context of the letter, it seems that Euler knew the exact form of (B.4) before his investigation of the Catalan numbers. Thus, once he found the product formula (B.1) he was able to derive (B.2) rather than simply guess it. We believe that Euler did not include his derivation in the first letter, as the generating function formula appears at the very end of it, but once Goldbach became interested he patiently explained all the steps, as well as other similar formulas.

B.3. Euler and Segner

Johann Andreas von Segner (1704–1777) was another frequent correspondent of Euler. Although Segner was older, Euler rose to prominence faster, and in 1755 crucially helped him to obtain a position at University of Halle; see [12, p. 40]. It seems there was a bit of competitive tension between them, which adversely affected the story.

In the late 1750s, Euler suggested to Segner the problem of counting the number of triangulations of an n-gon. We speculate,[2] based on Segner's later work, that Euler told him only values up to C_7, and neither the product formula (B.1) nor the generating function (B.2).

---

[2] Euler's letters to Segner did not survive, as Segner directed all his archives to be burned posthumously [21, p. 153]. Segner's letters to Euler did survive in St. Petersburg but have yet to be digitized (there are 159 letters between 1741 and 1771).

Segner accepted the challenge and in 1758 wrote a paper [61], whose main result is a recurrence relation which he finds and proves combinatorially:

$$C_{n+1} = C_0 C_n + C_1 C_{n-1} + C_2 C_{n-1} + \cdots + C_n C_0. \tag{B.5}$$

He then uses the formula to compute the values of C_n, $n \leq 18$, but makes an arithmetic mistake in computing $C_{13} = 742,900$, which then invalidates all larger values.

Euler must have realized that equation (B.5) is the last missing piece necessary to prove (B.3). He arranged for Segner's paper to be published in the journal of St. Petersburg Academy of Sciences, but with his own[3] *Summary* [20]. In it, he states (B.1), gives Segner a lavish compliment, then points out his numerical mistake, and correctly computes all C_n for $n \leq 23$. It seems unlikely that, given the simple product formula, Segner would have computed C_{13} incorrectly, so we assume that Euler shared it only with his close friend Goldbach, and kept Segner in the dark until after the publication.

In summary, a combination of results of Euler and Segner, combined with Goldbach's observation, gives a complete proof of the product formula (B.1). Unfortunately, it took about eighty years until the first complete proof was published.

B.4. Kotelnikow and Fuss

Semën Kirillovich Kotelnikow (1723–1806) was a Russian mathematician of humble origin who lived in St. Petersburg all his life. In 1766, soon after the Raid on Berlin, Euler returned to St. Petersburg. The same year Kotelnikow wrote a paper [38] elaborating on Catalan numbers. Although he claimed to have another way to verify (B.1), Larcombe notes that he "does little more than play around with the formula" [41].

Nicolaus Fuss (1755–1826) was a Swiss-born mathematician who moved to St. Petersburg to become Euler's assistant in 1773. He married Euler's granddaughter, became a well-known mathematician in his own right, and remained in Russia until death. In his 1795 paper [22], in response to Pfaff's question on the number of subdivisions of an n-gon into k-gons, Fuss introduced what are now known as the *Fuss–Catalan numbers* (see Problem A14), and gave a generalization of Segner's formula (B.5), but not the product formula.

---

[3] The article is unsigned, but the authorship by Euler is both evident and reported by numerous sources.

B.5. The French School, 1838–1843

In 1836, a young French mathematician Joseph Liouville (1809–1882) founded the *Journal de Mathématiques Pures et Appliquées*. He was in the center of mathematical life in Paris and maintained a large mailing list, which proved critical in this story.

In 1838, a Jewish French mathematician and mathematical historian Olry Terquem (1782–1862) asked Liouville if he knew a simple way to derive Euler's formula (B.1) from Segner's recurrence (B.5). Liouville in turn communicated this problem to "various geometers." What followed is a remarkable sequence of papers giving foundation to "Catalan studies."

First, Gabriel Lamé (1795–1870) wrote a letter to Liouville outlining the solution, a letter Liouville promptly published in the *Journal* and further popularized [39]. Lamé's solution was to use an elegant double counting argument. Let's count the number A_n of triangulations of an $(n + 2)$-gon with one of its $(n − 1)$ diagonals oriented. On the one hand, $A_n = 2(n − 1)C_n$. On the other hand, by summing over all possible directed diagonals we have

$$A_n = n\big(C_1 C_{n-1} + C_2 C_{n-2} + \cdots + C_n C_1\big).$$

Combining these two formulas with (B.5) easily implies (B.1).

In 1838, Belgian-born mathematician Eugène Charles Catalan (1814–1894) was a student of Liouville at École Polytechnique. Inspired by the work of Lamé, he became interested in the problem. He was the first to obtain what are now standard formulas

$$C_n = \frac{(2n)!}{n!(n+1)!} = \binom{2n}{n} - \binom{2n}{n-1}.$$

He then studied the problem of computing the number of different (non-associative) products of n variables, equivalent to counting the number of bracket sequences [14] (item 3).

Olinde Rodrigues (1795–1851) was a descendant of a large Sephardic Jewish family from Bordeaux. He received his doctorate in mathematics and had a career as a banker in Paris, but continued his mathematical interests. In the same volume of the *Journal*, he published two back-to-back short notes giving a more direct double counting proof of (B.1). The first note [58] gives a variation on Lamé's argument, a beautiful idea often regarded as folklore. Roughly, he counts in two ways the number B_n of triangulations of $(n + 2)$-gon where either an edge or a diagonal is oriented. On the one hand, $B_n = 2(2n + 1)C_{n-1}$. On the other hand, $B_n = (n + 2)C_{n+1}$, since triangulations with an oriented diagonal are in bijection with triangulations of an $(n + 3)$-gon obtained by inserting a triangle in place of a diagonal, and such edge can be any edge except the first one. We omit the details.

In [59] Rodrigues gives a related but even simpler argument for counting the bracket sequences. Denote by P_n the number of bracket sequences of labeled terms $x_1, \ldots, x_n$, e.g., $x_2(x_1x_3)$. Then on the one hand $P_n = n!C_{n-1}$. On the other hand, $P_{n+1} = (4n - 2)P_n$ since the variable x_{n+1} can be inserted into every bracket sequence in exactly $4n - 2$ ways. To see this, place a bracket around every variable and the whole product, e.g., $((x_2)((x_1)(x_3)))$. Now observe that a new variable is inserted immediately to the left of any of the $2n - 1$ left brackets, or immediately to the right of any of the $2n - 1$ right brackets, e.g.,

$$((x_2)((x_1)(x_3))) \rightarrow ((x_2)(((x_1)(x_4))(x_3))).$$

For further information on Rodrigues' contribution to Catalan numbers, see Tamm [67].

In 1839, clearly unaware of Euler's letters, a senior French mathematician Jacques Binet (1786–1856) wrote a paper [8] with a complete generating function proof of (B.1). Across the border in Germany, Johann August Grunert (1797–1872) became interested in the work of the Frenchmen on the one hand, and of Fuss on the other. A former student of Pfaff, in an 1841 paper [28] he found a product formula for the Fuss-Catalan numbers. He employed generating functions to reduce the problem to

$$Z(x)^m = \frac{Z(x) - 1}{x},$$

but seemed unable to finish the proof. The complete proof was given in 1843 by Liouville [44] using the *Lagrange inversion formula* (see [43]).[4]

After a few more mostly analytic papers inspired by the problem, the attention of the French school turned elsewhere. Catalan however returned to the problem on several occasions throughout his career. Even fifty years later, in 1878, he published *Sur les nombres de Segner* [16] on divisibility of the Catalan numbers.

B.6. The British School, 1857–1891

Rev. Thomas Kirkman (1806–1895) was a British ordained minister who had a strong interest in mathematics. In 1857, unaware of the previous work but with Cayley's support, he published a lengthy treatise [37]. There, he introduced the

---

[4] Liouville was clearly unaware of Fuss's paper until 1843, yet gives no credit to Grunert (cf. [42]).

Kirkman–Cayley numbers, defined as the number of ways to divide an n-gon with k nonintersecting diagonals (see Problem A42(d)), and states a general product formula, which he proves in a few special cases.

In 1857, Arthur Cayley (1821–1895) was a lawyer in London and extremely prolific mathematically. He was interested in a related counting of *plane trees*, and in 1859 published a short note [17], where (among other things) he gave a conventional generating function proof that the number of plane trees is the Catalan number. Like Kirkman, he was evidently unaware of the previous work. His final formula is

$$C_{m-1} = \frac{1 \cdot 3 \cdot 5 \cdots (2m-3)}{1 \cdot 2 \cdot 3 \cdots m} 2^{m-1},$$

which he called "a remarkably simple form." Curiously, in the same paper he also discovered and computed the generating function for the *ordered Bell numbers*. In 1860, Cayley won a professorship in Cambridge and soon became a central figure in British mathematics. A few years later Henry Martyn Taylor (1842–1927) became a student in Cambridge, where he remained much of his life, working in geometry, mathematical education, and politics.[5] In 1882, he and R. C. Rowe published a paper [68] where they carefully examined the literature of the French school but gave an erroneous description of the Euler-Segner story. They evidently missed later papers by Grunert and Liouville, and computed the Fuss-Catalan numbers along similar lines, using the Lagrange inversion formula; see [42].

Cayley continued exploring generating function methods for a variety of enumerative problems, among more than 900 papers he wrote. In 1891, Cayley used a generating function tour de force to completely resolve Kirkman's problem [18].

B.7. The Ballot Problem

The *ballot numbers* were first defined by Catalan in 1839, disguised as the number of certain triangulations [15]. We believe this contribution was largely forgotten since Catalan gave a formula for the ballot numbers in terms of the Catalan numbers, but neither gave a closed formula nor even a table of the first few values.

The ballot sequences were first introduced in 1878 by William Allen Whitworth (1840–1905), a British mathematician, a priest and a fellow at

---

[5] Despite his blindness, Taylor was elected a Mayor of Cambridge in 1900; see http://tinyurl.com/md99y9b.

Cambridge. He resolved the problem completely by an elegant counting argument and found interesting combinatorial applications [72]. Despite both geographical and mathematical proximity to Cayley, he did not notice that the numbers $1, 2, 5, 14, \ldots$ he computed are the Catalan numbers.

In modern terminology, the *ballot problem* was introduced in 1887 by Joseph Bertrand (1822–1900). In a half-page note [7], he defined the probability $P_{m,n}$ that in an election with $m + n$ voters, a candidate who gathered m votes is always leading the candidate with n votes. Bertrand announced that one can use induction to show that

$$P_{m,n} = \frac{m-n}{m+n}.$$

He famously concludes by saying:

> Il semble vraisemblable qu'un résultat aussi simple pourrait se démontrer d'une manière plus directe. (It seems probable that such a simple result could be shown in a more direct way.)

Within months, Bertrand's protégé Joseph-Émile Barbier (1839–1889) announced a generalization of $P_{m,n}$ to larger proportions of leading votes [5], a probabilistic version of the formula for the Fuss-Catalan numbers.

Désiré André (1840–1918) was a well-known French combinatorialist, a former student of Bertrand. He published an elegant proof of Bertrand's theorem [3] in the same 1887 volume of *Comptes Rendu* as Bertrand and Barbier. Although he is often credited with the *reflection method* or *reflection principle*, this attribution is incorrect,[6] as André's proof was essentially equivalent to that by Whitworth (cf. [56]). The reflection principle is in fact due to Dmitry Semionovitch Mirimanoff (1861–1945), a Russian-Swiss mathematician who discovered it in 1923, in a short note [50]. André's original proof was clarified and extended to Barbier's proposed generalization in [19].

We should mention here Takács's thorough treatment of the history of the ballot problem in probabilistic context [66], Humphreys's historical survey

---

[6] Marc Renault was first to note the mistake [56]; he traces the confusion to *Stochastic Processes* by J. L. Doob (1953) and *An Introduction to Probability Theory* by W. Feller (2nd ed., 1957). In a footnote, Feller writes: "The reflection principle is used frequently in various disguises, but without the geometrical interpretation it appears as an ingenious but incomprehensible trick. The probabilistic literature attributes it to D. André (1887). It appears in connection with the difference equations for random walks. These are related to some partial differential equations where the reflection principle is a familiar tool called *method of images*. It is generally attributed to Maxwell and Lord Kelvin."

on general lattice paths and the reflection principle [30], and Bru's historical explanation of how Bertrand learned it from Ampère [10].

B.8. Later Years

Despite a large literature, for decades the Catalan numbers remained largely unknown and unnamed in contrast with other celebrated sequences, such as the Fibonacci and Bernoulli numbers. Yet the number of publications on Catalan numbers was growing rapidly, so much that we cannot attempt to cover even a fraction of them. Here are few major appearances.

In 1870, Ernst Schröder (1841–1902) [60] published his famous list of four bracketing problems (with their solutions). The first problem is our item 3. For a short discussion of Schröder's four problems, see [65, pp. 177–178].

The first monograph citing the Catalan numbers is the *Théorie des Nombres* by Édouard Lucas (1842–1891), published in Paris in 1891. Despite the title, the book had a large combinatorial content, including Rodrigues's proof of Catalan's combinatorial interpretation in terms of brackets and products [45, p. 68].

The next notable monograph is the *Lehrbuch der Combinatorik* by Eugen Netto (1848–1919), published in Leipzig in 1901. This is one of the first combinatorics monographs; it includes a lot of material on permutations and combinations, and it devotes several sections to the papers by Catalan, Rodrigues, and some related work by Schröder [53, pp. 193–202].

We should mention that these monographs were more the exception than the rule. Many classical combinatorial books do not mention Catalan numbers at all, most notably Percy MacMahon, *Combinatorial Analysis* (1915/6), John Riordan, *An Introduction to Combinatorial Analysis* (1958), and H. J. Ryser, *Combinatorial Mathematics* (1963).

A crucial contribution was made by William G. Brown in 1965, when he recognized and collected a large number of references on the "Euler–Segner problem" and the "Pfaff–Fuss problem," as he called them [9].[7] From this point on, hundreds of papers involving Catalan numbers have been published and all standard textbooks began to include it (see, e.g., [29, §3.2] and [70, §3] for an early adoption). Some 465 references were assembled by Henry W. Gould in a remarkable bibliography [25], which first appeared in 1971 and was revised in 2007. At about that time, various lists of combinatorial interpretations of

---

[7] Unfortunately, Brown used somewhat confusing notation $D_{0,m}^{(3)}$ to denote the Catalan numbers.

Catalan numbers started to appear, see, e.g., [27, Appendix 1] and a very different kind of list in [69, p. 263].

B.9. The Name

Because of their chaotic history, the Catalan numbers have received this name relatively recently. In the old literature, they were sometimes called the *Segner numbers* or the *Euler-Segner numbers*, which is historically accurate as their articles were the first published work on the subject. Perhaps surprisingly, we are able to tell exactly who named them *Catalan numbers* and when.[8] Our investigation of this *eponymy* is informal, but we hope convincing (cf. [73] for an example of a proper historical investigation).

First, let us discard two popular theories: that the name was introduced by Netto in [53] or by Bell in [6]. Upon careful study of the text, it is clear that Netto did single out Catalan's work and in general was not particularly careful with the references,[9] but never specifically mentioned "Catalan Zahlen." Similarly, Eric Temple Bell (1883–1960) was a well-known mathematical historian, and referred to "Catalan's numbers" only in the context of Catalan's work. In a footnote, he in fact referred to them as the "Euler-Segner sequence" and clarifies the history of the problem.

Our investigation shows that the credit for naming Catalan numbers is due to an American combinatorialist John Riordan (1903–1988).[10] He tried this three times. The first two: *Math Reviews* MR0024411 (1948) and MR0164902 (1964) went unnoticed. Even Marshall Hall's influential 1967 monograph [29] does not have the name. But in 1968, when Riordan used "Catalan numbers" in the monograph [57], he clearly struck a chord.

Although Riordan's book is now viewed as somewhat disorganized and unnecessarily simplistic,[11] back in the day it was quite popular. It was lauded as "excellent and stimulating" in P. R. Stein's review, which continued to say "Combinatorial identities is, in fact, a book that must be read, from cover to cover, and several times." We are guessing it had a great influence on the field and cemented the terminology and some notation.

---

[8] This provides yet another example of the so-called *Stigler's Law of Eponymy*, that no result is named after its original discoverer.

[9] In fairness to Netto, this was standard at the time.

[10] Henry Gould's response seems to support this conclusion: http://tinyurl.com/mpyebyw.

[11] Riordan famously writes in the introduction to [57]: "Combinatorialists use recurrence, generating functions, and such transformations as the Vandermonde convolution; others, to my horror, use contour integrals, differential equations, and other resources of mathematical analysis."

We have further evidence of Riordan's authorship of a different nature: the *Ngram chart* for the words "Catalan numbers," which searches Google Books content for this phrase over the years.[12] The search clearly shows that the name spread after 1968 and that Riordan's monograph was the first book that contained it.

There were three more events on the way to the wide adoption of the name "Catalan numbers." In 1971, Henry Gould used it in the early version of [25], a bibliography of Catalan numbers with 243 entries. Then, in 1973, Neil Sloane gave this name to the sequence entry in [62], with Riordan's monograph and Gould's bibliography being two of the only five references. But the real popularity (as supported by the Ngram chart), was achieved after Martin Gardner's *Scientific American* column in June 1976, popularizing the subject:

> [Catalan numbers] have [...] delightful propensity for popping up unexpectedly, particularly in combinatorial problems. Indeed, the Catalan sequence is probably the most frequently encountered sequence that is still obscure enough to cause mathematicians lacking access to N. J. A. Sloane's *Handbook of Integer Sequences* to expend inordinate amounts of energy re-discovering formulas that were worked out long ago. [23]

B.10. The Importance

Now that we have covered more than 200 years of the history of Catalan numbers, it is worth pondering if the subject matter is worth the effort. Here we defer to others, and include the following helpful quotes, in no particular order.

Manuel Kauers and Peter Paule write in their 2011 monograph:

> It is not exaggerated to say that the Catalan numbers are the most prominent sequence in combinatorics. [31]

Peter Cameron, in his 2013 lecture notes, elaborates:

> The Catalan numbers are one of the most important sequences of combinatorial numbers, with a large range of occurrences in apparently different counting problems. [13]

Martin Aigner gives a slightly different emphasis:

> The Catalan numbers are, next to the binomial coefficients, the best studied of all combinatorial counting numbers. [2]

---

[12] This chart is available here: `http://tinyurl.com/k9nvf28`.

This flattering comparison was also mentioned by Neil Sloane and Simon Plouffe in 1995, in the final print edition of EIS:

> Catalan numbers are probably the most frequently occurring combinatorial numbers after the binomial coefficients. [63]

In the OEIS itself, in regard to the Catalan number sequence A000108, appears the statement:

> This is probably the longest entry in the OEIS, and rightly so.

Doron Zeilberger attempts to explain this in his *Opinion 49*:

> Mathematicians are often amazed that certain mathematical objects (numbers, sequences, etc.) show up so often. For example, in enumerative combinatorics we encounter the Fibonacci and Catalan sequences in many problems that seem to have nothing to do with each other. [...] The answer, once again, is our human predilection for triviality. [...] The Catalan sequence is the simplest sequence whose generating function is a (genuine) algebraic formal power series. [74]

In a popular article, Jon McCammond reveals a new side of Catalan numbers:

> The Catalan numbers are a favorite pastime of many amateur (and professional) mathematicians. [49]

Christian Aebi and Grant Cairns give both a praise and a diagnosis:

> Catalan numbers are the subject of [much] interest (sometimes known as *Catalan disease*). [1]

Thomas Koshy, in his introductory book on Catalan numbers, gives them literally a heavenly praise:

> Catalan numbers are even more fascinating [than the *Fibonacci numbers*]. Like the North Star in the evening sky, they are a beautiful and bright light in the mathematical heavens. They continue to provide a fertile ground for number theorists, especially, Catalan enthusiasts and computer scientists. [35]

In conclusion, let us mention the following answer which Richard Stanley gave in a 2008 interview on how the Catalan numbers exercise came about. See also the preface of the present monograph.

> I'd have to say my favorite number sequence is the Catalan numbers. [...] Catalan numbers just come up so many times. It was well-known before me that they had many different combinatorial interpretations. [...] When I started teaching enumerative combinatorics, of course I did the Catalan numbers. When I started

doing these very basic interpretations – any enumerative course would have some of this – I just liked collecting more and more of them and I decided to be systematic. Before, it was just a typed up list. When I wrote the book, I threw everything I knew in the book. Then I continued from there with a website, adding more and more problems. [36]

Acknowledgments: We are very grateful to Richard Stanley for the suggestion to write this appendix. We would also like to thank Maria Rybakova for her help with translation of [20], to Xavier Viennot for showing us the scans of Euler's original letter to Goldbach, and to Peter Larcombe for sending me his paper and help with other references. The author was partially supported by the NSF.

Glossary

alternating permutation $a_1 a_2 \cdots a_n \in \mathfrak{S}_n$: $a_1 > a_2 < a_3 > a_4 < \cdots$. See also "Euler number."

alternating tree: an (unrooted) tree with n vertices labeled $1, 2, \ldots, n$ for which every vertex is either less than all its neighbors or greater than all its neighbors.

antichain (of a poset P): a subset of P for which no two elements are comparable.

ascending run (of a permutation w): a maximal increasing sequence of consecutive terms of w.

ascent (of a permutation $a_1 a_2 \cdots a_n \in \mathfrak{S}_n$): an integer $1 \leq i \leq n - 1$ satisfying $a_i < a_{i+1}$.

avoiding permutation: if $w = a_1 \cdots a_k$ is a permutation of $1, 2, \ldots, k$, then a permutation $v = b_1 \cdots b_n \in \mathfrak{S}_n$ *avoids* w if no subsequence $b_{i_1} \cdots b_{i_k}$ of v has its elements in the same relative order as w, i.e., $a_r < a_s$ if and only if $b_{i_r} < b_{i_s}$ for all $1 \leq r < s \leq k$.

Baxter permutation: a permutation $w \in \mathfrak{S}_n$ satisfying: if $w(r) = i$ and $w(s) = i + 1$, then there is a k_i between r and s (i.e., $r \leq k_i \leq s$ or $s \leq k_i \leq r$) such that $w(t) \leq i$ if t is between r and k_i, while $w(t) \geq i + 1$ if t is between $k_i + 1$ and s. For instance, all permutations $w \in \mathfrak{S}_4$ are Baxter permutations except 2413 and 3142.

bijection: a function $f \colon X \to Y$ from a set X to a set Y that is both injective (one-to-one) and surjective (onto).

boolean algebra (as a type of finite poset): a poset isomorphic to the set of all subsets of a finite set, ordered by inclusion.

191

Bruhat order (on $\mathfrak{S}_n$): the partial ordering on the symmetric group $\mathfrak{S}_n$ such that w covers v if $w = v(i,j)$ and $\mathrm{inv}(w) = 1 + \mathrm{inv}(v)$, where (i,j) is the transposition interchanging i and j, and inv denotes the number of inversions.

chain: a poset P such that if $s,t \in P$ then either $s \leq t$ or $s > t$. The poset with elements $1, 2, \ldots, n$ in their usual order is a chain denoted $\boldsymbol{n}$.

code (of a partition λ): let D_λ denote the Young diagram of λ, with its left-hand edge and upper edge extended to infinity. (Hopefully the figure below will make this description and other parts of this definition clear.) Put a 0 next to each vertical edge of the "lower boundary" of D_λ, and a 1 next to each horizontal edge. If we read these numbers as we move north and east along the lower boundary, then we obtain an infinite binary sequence $C_\lambda = \cdots c_{-2} c_{-1} c_0 c_1 c_2 \cdots$, indexed so that the first 1 in the sequence is $c_{1-\ell}$, where ℓ is the number of parts of λ. (This indexing is irrelevant for the use of C_λ in this monograph but is important for other purposes.) We call C_λ the *code* of λ. The figure below shows that

$$C_{441} = \cdots 0001011100111 \cdots,$$

where the first 1 is $c_{1-3} = c_{-2}$.

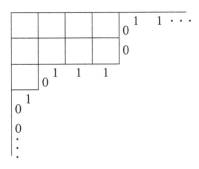

column-strict plane partition: a plane partition for which the nonzero entries in each column are strictly decreasing.

(complete) flag variety: the set of all maximal chains $0 = V_0 \subset V_1 \subset \cdots \subset V_n = V$ of subspaces of an n-dimensional vector space V over a field K. It has the natural structure of an algebraic variety.

complete matching (of a graph G with no isolated vertices): a set M of edges of G such that every vertex of G is incident to exactly one of the edges in M. Thus if $\#M = m$, then G has $2m$ vertices.

composition (of an integer $n \geq 0$): a sequence $(a_1, a_2, \ldots, a_k)$ of positive integers satisfying $\sum a_i = n$.

conjugate partition (to an integer partition λ): the partition whose diagram is the transpose (reflection through the main diagonal) to that of λ.

convex subset (of a poset P): a subset C of P such that if $s, u \in C$ and $s < t < u$, then $t \in C$.

core: a partition λ of an integer is a *p-core* if it has no hook lengths equal to (equivalently, divisible by) p.

cover: in a poset P, we say that v *covers* u (or u is *covered by* v) if $u < v$ and no $w \in P$ satisfies $u < w < v$.

descent set (of a permutation $a_1 a_2 \cdots a_n \in \mathfrak{S}_n$): $D(w) = \{1 \le i \le n - 1 : a_i > a_{i+1}\}$.

difference operator Δ: the operator defined by $\Delta f(n) = f(n+1) - f(n)$, where f is any function from $\mathbb{Z}$ to an abelian group. Notation like $\Delta^k f(0)$ is short for $\Delta^k f(n)\big|_{n=0}$.

direct sum decomposition: see "poset of direct-sum decompositions."

doubly stochastic matrix: a square matrix of nonnegative real numbers for which every row and column sums to 1.

Durfee square (of the Young diagram of a partition λ): the largest square subdiagram that contains the upper left-hand corner.

Ehrhart polynomial: see "Ehrhart quasipolynomial."

Ehrhart quasipolynomial (of a convex polytope $\mathcal{P}$ in $\mathbb{R}^d$ with rational vertices): the function $i(\mathcal{P}, n)$ defined for $n \in \mathbb{P}$ by

$$i(\mathcal{P}, n) = \#(n\mathcal{P} \cap \mathbb{Z}^d),$$

where $n\mathcal{P} = \{nv : v \in \mathcal{P}\}$. A basic theorem states that $i(\mathcal{P}, n)$ is a quasipolynomial. If $\mathcal{P}$ has integer vertices, then $i(\mathcal{P}, n)$ is a polynomial in n, called the **Ehrhart polynomial** of $\mathcal{P}$.

Eulerian poset: a (finite) poset with a unique minimal element $\hat{0}$ and unique maximal element $\hat{1}$, such that all maximal chains have the same length and every interval $[s, t]$ with $s < t$ has the same number of elements of odd rank as elements of even rank.

Euler number E_n: the number of alternating permutations $w = a_1 a_2 \cdots a_n$ in $\mathfrak{S}_n$, i.e., $a_1 > a_2 < a_3 > a_4 < a_5 > \cdots$. Equivalently,

$$\sum_{n \ge 0} E_n \frac{x^n}{n!} = \sec x + \tan x.$$

exponential formula: given a function $f: \mathbb{P} \to \mathbb{R}$, define a new function $h: \mathbb{N} \to \mathbb{R}$ by

$$h(0) = 1$$

$$h(n) = \sum_{\pi = \{B_1, \ldots, B_k\} \in \Pi_n} f(\#B_1) \cdots f(\#B_k), \ n > 0,$$

where Π_n denotes the set (or lattice) of all partitions of $[n]$. Then

$$\sum_{n \geq 0} h(n) \frac{x^n}{n!} = \exp \sum_{n \geq 1} f(n) \frac{x^n}{n!}.$$

flag f-vector (of a graded poset P of rank $n + 1$ with $\hat{0}$ and $\hat{1}$): the function $\alpha_P: 2^{[n]} \to \mathbb{Z}$, where $\alpha_P(S)$ is the number of chains $\hat{0} = t_0 < t_1 < \cdots < t_{k-1} < t_k = \hat{1}$ of P such that $S = \{\rho(t_1), \ldots, \rho(t_{k-1})\}$. Here ρ is the rank function of P, $\hat{0}$ is the unique minimal element of P, and $\hat{1}$ is the unique maximal element of P.

flag variety: see "(complete) flag variety."

graded poset of rank m: a (finite) poset P for which all maximal chains have length m. There is then a unique *rank function* $\rho: P \to \mathbb{Z}$ satisfying $\rho(t) = 0$ if t is a minimal element of P, $\rho(t) = m$ if t is a maximal element of P, and $\rho(t) = \rho(s) + 1$ if t covers s in P.

Hadamard product (of two power series $\sum a_n x^n$ and $\sum b_n x^n$): the power series $\sum a_n b_n x^n$.

Hasse diagram (of a finite poset P): a graph drawn in the plane whose vertices are the elements of P, with t placed higher than s if $s < t$, and with an (undirected) edge between two vertices s and t if s covers t or t covers s.

hook lengths (of a partition λ): the multiset of numbers $h(u)$, where u is a square of the Young diagram of λ and $h(u)$ is the number of squares directly to the right of u or directly below u, containing u itself once. The partition $(3, 2)$ has hook lengths (in the usual reading order of the squares of the Young diagram) 4,3,1,2,1.

horizontally convex polyomino: a polyomino P for which each "row" is an unbroken line of squares, that is, if L is any line segment parallel to the x-axis whose two endpoints are in P, then $L \subset P$.

i-Eulerian polynomial: the polynomial $(1 - x)^{d+1} \sum_{n \geq 0} i(n) x^n$, where $i(n)$ is a polynomial of degree d.

incomparability graph (of a poset P): the graph whose vertices are the elements of P, with vertices u and v adjacent if they are incomparable in P (i.e., neither $u \leq v$ nor $v < u$).

increasing tree: a (rooted) tree with n vertices labeled $1, 2, \ldots, n$ such that the vertices increase along any path from the root.

indecomposable permutation: a permutation $w = a_1 a_2 \cdots a_n \in \mathfrak{S}_n$ such that $\{a_1, a_2, \ldots, a_i\} \neq \{1, 2, \ldots, i\}$ for $1 \leq i \leq n-1$.

induced subposet (of a poset P): a poset Q whose elements are a subset of the elements of P, with $s < t$ in Q if and only if $s, t \in Q$ and $s < t$ in P.

induced subtree (of a graph G): a tree T whose vertices are a subset of the vertices of G, with u, v adjacent in T if and only if u, v are vertices of T which are adjacent in G.

inversion (of a permutation $w = w_1 \cdots w_n \in \mathfrak{S}_n$): a pair (w_i, w_j), where $i < j$ and $w_i > w_j$.

inversion table (of a permutation $w = w_1 \cdots w_n \in \mathfrak{S}_n$): the sequence $(a_1, \ldots, a_n)$, where a_i is the number of entries j of w to the left of i satisfying $j > i$.

involution: a permutation $w \in \mathfrak{S}_n$ satisfying $w^2 = 1$. Equivalently, all cycles of w have length one or two.

Kummer's theorem (on binomial coefficients): if $0 \leq k \leq n$ and p is prime, then the exponent of the largest power of p dividing $\binom{n}{k}$ is equal to the number of carries in adding k and $n - k$ in base p (using the usual addition algorithm).

Lagrange inversion formula: Let $F(x) = a_1 x + a_2 x^2 + \cdots$ be a formal power series (over $\mathbb{C}$, say) with $a_1 \neq 0$, and let $F^{\langle -1 \rangle}$ denote its compositional inverse, i.e., $F(F^{\langle -1 \rangle}(x)) = F^{\langle -1 \rangle}(F(x)) = x$. Write $[x^n]G(x)$ for the coefficient of x^n in the power series $G(x)$. Then for $k, n \in \mathbb{Z}$,

$$n[x^n]F^{\langle -1 \rangle}(x)^k = k[x^{n-k}]\left(\frac{x}{F(x)}\right)^n.$$

lattice (as a type of poset): a poset for which any two elements s, t have a least upper bound (or join) and greatest lower bound (or meet).

leaf (of a tree): a vertex with no children.

left-to-right maximum (of a permutation $a_1 a_2 \cdots a_n \in \mathfrak{S}_n$): a term a_j such that $a_j > a_i$ for all $i < j$.

length (of a chain C in a finite poset): $\#C - 1$.

linear extension (of a p-element poset P): a bijection $f \colon P \to [p]$ such that if $s < t$ in P, then $f(s) < f(t)$. If $P = \{t_1, \ldots, t_p\}$, then f is sometimes represented by the permutation $f^{-1}(1), f^{-1}(2), \ldots, f^{-1}(p)$ of the elements of P.

matroid: a collection $\mathcal{I}$ of subsets, called *independent sets*, of a (finite) set X such that (a) if $I \in \mathcal{I}$ and $I' \subset I$, then $I' \in \mathcal{I}$; and (b) for any $Y \subseteq X$, the maximal independent sets $I \subseteq Y$ all have the same cardinality. The maximal independent sets of X are called *bases*. (We have given one of many equivalent definitions of a matroid.)

meet semilattice: a poset L for which any two elements $s, t \in L$ have a greatest lower bound.

Möbius function (of a finite poset P): the function $\mu \colon \mathrm{Int}(P) \to \mathbb{Z}$, where $\mathrm{Int}(P)$ denotes the set of (nonempty) closed intervals $[s, t]$ of P, which is defined recursively by the conditions

$$\mu(t, t) = 1, \quad \text{for all } t \in P$$

$$\sum_{u \in [s,t]} \mu(s, u) = 0, \quad \text{for all } s < t \text{ in } P.$$

Möbius inversion (for finite posets P). The *Möbius inversion formula* states that the following two conditions are equivalent on functions $f, g \colon P \to A$, where A is an additive abelian group (such as $\mathbb{Z}$):

$$g(t) = \sum_{s \leq t} f(s), \quad \text{for all } t \in P$$

$$f(t) = \sum_{s \leq t} g(s) \mu(s, t), \quad \text{for all } t \in P,$$

where μ denotes the Möbius function of P.

multiset: informally, a set with repeated elements such as $\{1, 1, 1, 3, 4, 4\}$, written more succinctly as $\{1^3, 3, 4^2\}$. If the elements of a multiset M all belong to a set S, then we say that M is a multiset *on* S. (We don't require that every element of S appears in M.) A more formal definition of multiset can easily be given but is not needed here.

natural partial ordering (of $[n]$): a partial ordering P of $[n]$ such that if $i <_P j$ then $i <_{\mathbb{Z}} j$.

ordered Bell number: the number $f(n)$ of ordered set partitions of an n-element set S, that is, partitions of S whose blocks are linearly ordered. Cayley showed that $\sum_{n \geq 0} f(n) \frac{x^n}{n!} = (2 - e^x)^{-1}$.

order ideal (of a poset P): a subset I of P such that if $t \in I$ and $s < t$ in P, then $s \in I$. The poset (actually a distributive lattice) of all order ideals of P, ordered by inclusion, is denoted $J(P)$.

ordinal sum (of two disjoint posets P and Q): the poset $P \oplus Q$ on the disjoint union of the elements of P and Q such that $s \leq t$ in $P \oplus Q$ if either (a) $s \leq t$ in P, or (b) $s \leq t$ in Q, or (c) $s \in P, t \in Q$.

partially ordered set: a set P with a binary operation $\leq$ satisfying the following axioms: (a) $t \leq t$ for all $t \in P$, (b) if $s \leq t$ and $t \leq s$, then $s = t$, and (c) if $s \leq t$ and $t \leq u$, then $s \leq u$. The notation $s < t$ means that $s \leq t$ and $s \neq t$, while $t \geq s$ means that $s \leq t$.

partition (of a finite set S): a collection $\{B_1, \ldots, B_k\}$ of nonempty pairwise disjoint subsets (called *blocks*) of S whose union is S. The equivalence classes of an equivalence relation on S form a partition of S, and conversely, every partition of S is the set of equivalence classes of an equivalence relation on S. The set of all partitions of $[n]$ is denoted Π_n.

partition (of an integer $n \geq 0$): a sequence $(\lambda_1, \lambda_2, \ldots)$ of integers satisfying $\lambda_1 \geq \lambda_2 \geq \cdots \geq 0$ and $\sum \lambda_i = n$. Often trailing 0's are omitted from the notation, so the five partitions of 4 are (using abbreviated notation) 4, 31, 22, 211, 1111. See [64, p. 58 and §1.8].

peak (of a Dyck, Motzkin, or Schröder path): an up step followed by a down step.

plane partition (of n): an array $\pi = (\pi_{ij})_{i,j \geq 1}$ of nonnegative integers such that every row and column is weakly decreasing and $\sum \pi_{ij} = n$. The six plane partitions of 3 are (omitting entries equal to 0)

$$
\begin{array}{cccccc}
3 & 2\,1 & 2 & 1\,1\,1 & 1\,1 & 1 \\
  &   & 1 &   & 1   & 1 \; . \\
  &   &   &   &   & 1
\end{array}
$$

polyomino: a finite union P of unit squares in the plane such that the vertices of the squares have integer coordinates, and P is connected and has no finite cut set. We consider here two polyominos to be *equivalent* if there is a translation that transforms one into the other.

poset: short for partially ordered set.

poset of direct sum decompositions (of an n-dimensional vector space V over a field K): the poset whose elements are sets $\{V_1, \ldots, V_k\}$ of subspaces of V

such that $\dim V_i > 0$ and $V = V_1 \oplus \cdots \oplus V_k$ (vector space direct sum). Define $\{U_1, \ldots, U_j\} \leq \{V_1, \ldots, V_k\}$ if each U_i is a subspace of some V_h.

q-binomial coefficient $\binom{n}{k}$ (for integers $0 \leq k \leq n$): the expression

$$\frac{(q^n - 1)(q^{n-1} - 1) \cdots (q^{n-k+1} - 1)}{(q^k - 1)(q^{k-1} - 1) \cdots (q - 1)}.$$

It is not difficult to show that $\binom{n}{k}$ is a polynomial in q with nonnegative integer coefficients, whose value at $q = 1$ is the binomial coefficient $\binom{n}{k}$.

quasipolynomial: a function $f \colon \mathbb{Z} \to \mathbb{C}$ (or more generally $\mathbb{Z} \to A$ where A is a commutative ring with identity) of the form $c_d(n)n^d + c_{d-1}(n)n^{d-1} + \cdots + c_0(n)$, where each $c_i(n)$ is periodic.

quasisymmetric function (in the variables $x_1, \ldots, x_n$): a polynomial in $x_1, \ldots, x_n$ such that for all sequences $1 \leq i_1 < i_2 < \cdots < i_k \leq n$ and all sequences $a_1, \ldots, a_k$ of positive integers, the coefficient of $x_{i_1}^{a_1} \cdots x_{i_k}^{a_k}$ is equal to the coefficient of $x_1^{a_1} \cdots x_k^{a_k}$.

rank-generating function (of a finite graded poset P): the polynomial $F(P, q) = \sum_i p_i q^i$, where P has p_i elements of rank i.

relative volume (of a d-dimensional convex polytope $\mathcal{P}$ in $\mathbb{R}^n$ with integer vertices): Let $\mathcal{A}$ denote the affine span of $\mathcal{P}$ and $L = \mathcal{A} \cap \mathbb{Z}^n$. Let $\varphi \colon \mathcal{A} \to \mathbb{R}^d$ be a bijective linear map that takes L to $\mathbb{Z}^d$. The relative volume of $\mathcal{P}$ is then the ordinary d-dimensional volume of $\varphi(\mathcal{P})$. For instance, the relative volume of the line segment connecting $(0,0)$ to $(2,4)$ is 2.

restricted growth function: A sequence $a_1, \ldots, a_n$ of positive integers such that if $k > 1$ appears in the sequence, then the first occurrence of $k - 1$ appears before the first occurrence of k.

reverse alternating permutation $a_1 a_2 \cdots a_n \in \mathfrak{S}_n$: $a_1 < a_2 > a_3 < a_4 > \cdots$.

RSK algorithm: a certain bijection φ (not defined here) between the symmetric group $\mathfrak{S}_n$ and pairs (P, Q) of standard Young tableaux of the same shape $\lambda \vdash n$. Two properties of RSK are the following: (a) the length of the longest increasing subsequence of w is equal to λ_1, the largest part of λ, and (b) w is an involution if and only if $P = Q$, in which case the number of fixed points of w is equal to the number of odd parts of the conjugate partition λ'.

semilattice. See "meet semilattice."

shuffle (of two sequences α and β): a sequence that contains a subsequence identical to α, and when we remove this subsequence we obtain β.

simple graph: a graph with no loops (an edge from a vertex to itself) and multiple edges (more than one edge incident to the same two vertices).

simplicial poset: a (finite) poset P with a unique minimal element $\hat{0}$ such that every interval $[\hat{0}, t]$ is isomorphic to a boolean algebra.

Smith normal form (of an $n \times n$ matrix A over a commutative ring with 1): a diagonal matrix of the form $D = \mathrm{diag}(e_1, e_2, \ldots, e_n) = PAQ$ where P, Q are invertible (over R) $n \times n$ matrices with entries in R, and where $e_i \in R$ and e_{i+1} divides e_i in R for $2 \leq i \leq n$. If R is a principal ideal domain (PID), then the Smith normal form always exists and is unique up to multiplication by units in R. If R is not a PID, such as $\mathbb{Z}[q]$, then in general a Smith normal form does not exist, though it may exist in special cases.

standard Young tableau (SYT) of shape $\lambda = (\lambda_1, \lambda_2, \ldots)$, where λ is a partition of n: a left-justified array containing the numbers $1, 2, \ldots, n$ once each, with λ_i numbers in the ith row, and where every row and column is increasing. An example of an SYT of shape $(5, 2, 2)$ is given by

$$
\begin{array}{ccccc}
1 & 3 & 4 & 7 & 9 \\
2 & 5 & & & \\
6 & 8 & & &
\end{array} \; .
$$

SYT: short for standard Young tableau.

weak (Bruhat) order (on $\mathfrak{S}_n$): the partial ordering on the symmetric group $\mathfrak{S}_n$ whose cover relations are

$$
a_1 a_2 \cdots a_i a_{i+1} \cdots a_n \lessdot a_1 a_2 \cdots a_{i+1} a_i \cdots a_n
$$

whenever $a_i < a_{i+1}$. For instance, 526314 covers 256314, 523614, and 526134.

weak ordered partition (of a finite set S): a sequence $(S_1, \ldots, S_k)$ of pairwise disjoint sets whose union is S. "Weak" means that we allow $S_i = \emptyset$.

winding number (of a directed closed curve C in $\mathbb{R}^2$ with respect to a point p not on C): intuitively, the number of times C wraps around p in a counterclockwise direction as we traverse C in the direction that it is directed. A precise definition may be found, for instance, at Wikipedia.

Young diagram (of an integer partition $\lambda = (\lambda_1, \lambda_2, \ldots)$): a left-justified array of squares with λ_i squares in the ith row. For instance, the Young diagram of the partition $(5, 2, 2)$ looks like

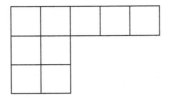

Young's lattice: the poset (actually a distributive lattice) of all partitions of all integers $n \geq 0$, ordered componentwise, i.e., $(\lambda_1, \lambda_2, \ldots) \leq (\mu_1, \mu_2, \ldots)$ if $\lambda_i \leq \mu_i$ for all $i \geq 1$.

Bibliography

[1] C. Aebi and G. Cairns, Catalan numbers, primes, and twin primes, *Elem. Math.* **63** (2008), 153–164.

[2] M. Aigner, Catalan and other numbers, in *Algebraic Combinatorics and Computer Science*, Springer, Berlin, 2001, 347–390.

[3] D. André, Solution directe du problème résolu par M. Bertrand, *Comptes Rendus Acad. Sci. Paris* **105** (1887), 436–437.

[4] G. E. Andrews, *The Theory of Partitions*, Addison-Wesley, Reading, MA, 1976.

[5] É. Barbier, Généralisation du problème résolu par M. J. Bertrand, *Comptes Rendus Acad. Sci. Paris* **105** (1887), 407.

[6] E. T. Bell, The iterated exponential integers, *Annals of Math.* **39** (1938), 539–557.

[7] J. Bertrand, Solution d'une problème, *Comptes Rendus Acad. Sci. Paris* **105** (1887), 369.

[8] M. J. Binet, Réflexions sur le problème de déterminer le nombre de manières dont une figure rectiligne peut être partagées en triangles au moyen de ses diagonales, *J. Math. Pures Appl.* **4** (1839), 79–90.

[9] W. G. Brown, Historical note on a recurrent combinatorial problem, *Amer. Math. Monthly* **72** (1965), 973–979.

[10] B. Bru, Les leçons de calcul des probabilités de Joseph Bertrand, *J. Électron. Hist. Probab. Stat.* **2** (2006), no. 2, 44 pp.

[11] N. G. de Bruijn and B. J. M. Morselt, A note on plane trees, *J. Combinatorial Theory* **2** (1967), 27–34.

[12] R. S. Calinger, Leonhard Euler: life and thought, in *Leonhard Euler: Life, Work and Legacy* (R. E. Bradley and E. Sandifer, editors), Elsevier, Amsterdam, 2007.

[13] P. J. Cameron, *LTCC course notes on Enumerative Combinatorics*, Lecture 3: Catalan numbers (Autumn 2013); available at http://www.maths.qmul.ac.uk/~pjc/ec/.

[14] E. C. Catalan, Note sur une équation aux différences finies, *J. Math. pure et appliquées* **3** (1838), 508–516.

[15] E. C. Catalan, Solution nouvelle de cette question: un polygone étant donné, de combien de manières peut-on le partager en triangles au moyen de diagonales?, *J. Math. Pures Appl.* **4** (1839), 91–94.

[16] E. Catalan, Sur les nombres de Segner, *Rend. Circ. Mat. Palermo* **1** (1887), 190–201.

[17] A. Cayley, On the analytical forms called trees II, *Philos. Mag.* **18** (1859), 374–378.

[18] A. Cayley, On the partitions of a polygon, *Proc. London Math. Soc.* **22** (1891), 237–262.

[19] A. Dvoretzky and Th. Motzkin, A problem of arrangements, *Duke Math. J.* **14** (1947), 305–313.

[20] L. Euler, Summary of [61] in the same volume of *Novi Commentarii*, 13–15.

[21] E. A. Fellmann, *Leonhard Euler*, Birkhäuser, Basel, 2007.

[22] N. Fuss, Solutio quaestionis, quot modis polygonum *n* laterum in polygona *m* laterum, per diagonales resolvi queat, *Novi Acta Acad. Sci. Petrop.* **9** (1795), 243–251; available at [55].

[23] M. Gardner, Mathematical Games, Catalan numbers: an integer sequence that materializes in unexpected places, *Scientific Amer.* **234**, no. 6 (June 1976), 120–125, 132.

[24] I. M. Gelfand, M. M. Kapranov, and A. V. Zelevinsky, *Discriminants, Resultants and Multidimensional Determinants*, Birkhäuser Boston, Cambridge, Massachusetts, 1994.

[25] H. W. Gould, *Bell and Catalan Numbers: A Research Bibliography of Two Special Number Sequences*, Revised 2007 edition; available at `http://tinyurl.com/opkn1h8`.

[26] I. P. Goulden and D. M. Jackson, *Combinatorial Enumeration*, John Wiley, New York, 1983; reissued by Dover, New York, 2004.

[27] J.A.S. Growney, Finitely generated free groupoids, PhD thesis, University of Oklahoma, 1970.

[28] J. A. Grunert, Ueber die Bestimmung der Anzahl der verschiedenen Arten, auf welche sich ein *n*eck durch Diagonales in lauter *m*ecke zerlegen lässt, mit Bezug auf einige Abhandlungen der Herren Lamé, Rodrigues, Binet, Catalan und Duhamel in dem Journal de Mathématiques Pure et Appliquées, publié par Joseph Liouville, Vols. 3, 4, *Arch. Math. Physik* **1** (1841), 193–203.

[29] M. Hall, *Combinatorial Theory*, Blaisdell, Waltham, Massachusetts, 1967.

[30] K. Humphreys, A history and a survey of lattice path enumeration, *J. Statist. Plann. Inference* **140** (2010), 2237–2254.

[31] M. Kauers and P. Paule, *The Concrete Tetrahedron*, Springer, Vienna, 2011.

[32] D. A. Klarner, Correspondences between plane trees and binary sequences, *J. Combinatorial Theory* **9** (1970), 401–411.

[33] D. E. Knuth, *The Art of Computer Programming*, vol. 1, *Fundamental Algorithms*, Addison-Wesley, Reading, Massachusetts, 1968; second ed., 1973.

[34] D. E. Knuth, *The Art of Computer Programming*, vol. 3, 2nd ed., Addison-Wesley, Reading, Massachusetts, 1998.

[35] T. Koshy, *Catalan Numbers with Applications*, Oxford University Press, Oxford, 2009.

[36] H. S. Kim, Interview with Professor Stanley, *Math Majors Magazine*, vol. 1, no. 1 (December 2008), pp. 28–33; available at `http://tinyurl.com/q423c61`.

[37] T. P. Kirkman, On the *K*-Partitions of the *R*-Gon and *R*-Ace, *Phil. Trans. Royal Soc.* **147** (1857), 217–272.

[38] S. Kotelnikow, Demonstatio seriei $\frac{4 \cdot 6 \cdot 10 \cdot 14 \cdot 18 \cdot 22 \cdots (4n-10)}{2 \cdot 3 \cdot 4 \cdot 5 \cdot 6 \cdot 7 \cdots (n-1)}$ exhibitae in recensione VI. tomi VII. Commentariorum A. S. P., *Novi Comment. Acad. Sci. Imp. Petropol.* **10** (1766), 199–204; available at [55].

[39] G. Lamé, Extrait d'une lettre de M. Lamé à M. Liouville sur cette question: Un polygone convexe étant donné, de combien de manières peut-on le partager en triangles au moyen de diagonales?, *Journal de Mathématiques pure et appliquées* **3** (1838), 505–507.

[40] P. J. Larcombe, The 18th century Chinese discovery of the Catalan numbers, *Math. Spectrum* **32** (1999/2000), 5–7.

[41] P. J. Larcombe, On pre-Catalan Catalan numbers: Kotelnikow (1766), *Math. Today* **35** (1999), no. 1, 25.

[42] P. J. Larcombe and P.D.C. Wilson, On the trail of the Catalan sequence, *Mathematics Today* **34** (1998), 114–117.

[43] P. J. Larcombe and P.D.C. Wilson, On the generating function of the Catalan sequence, *Congr. Numer.* **149** (2001), 97–108.

[44] J. Liouville, Remarques sur un mémoire de N. Fuss, *J. Math. Pures Appl.* **8** (1843), 391–394.

[45] E. Lucas, *Théorie des Nombres*, Gauthier-Villard, Paris, 1891.

[46] J. J. Luo, Antu Ming, the first inventor of Catalan numbers in the world, *Neimengu Daxue Xuebao* **19** (1988), 239–245 (in Chinese).

[47] J. Luo, Ming Antu and his power series expansions, in *Seki, founder of modern mathematics in Japan*, Springer, Tokyo, 2013, 299–310.

[48] P. A. MacMahon, *Combinatory Analysis*, vols. 1 and 2, Cambridge University Press, 1916; reprinted by Chelsea, New York, 1960, and by Dover, New York, 2004.

[49] J. McCammond, Noncrossing partitions in surprising locations, *Amer. Math. Monthly* **113** (2006), 598–610.

[50] D. Mirimanoff, A propos de l'interprétation géométrique du problème du scrutin, *L'Enseignement Math.* **23** (1923), 187–189.

[51] S. G. Mohanty, *Lattice Path Counting and Applications*, Academic Press, New York, 1979.

[52] T. V. Narayana, *Lattice Path Combinatorics with Statistical Applications*, Mathematical Expositions no. 23, University of Toronto Press, Toronto, 1979.

[53] E. Netto, *Lehrbuch der Combinatorik*, Teubner, Leipzig, 1901.

[54] I. Pak, Who computed Catalan numbers? (February 20, 2013), Who named Catalan numbers? (February 5, 2014), blog posts on *Igor Pak's blog*; available at http://igorpak.wordpress.com/.

[55] I. Pak, *Catalan Numbers website*, http://www.math.ucla.edu/~pak/lectures/Cat/pakcat.htm.

[56] M. Renault, Lost (and found) in translation, *Amer. Math. Monthly* **115** (2008), 358–363.

[57] J. Riordan, *Combinatorial Identities*, John Wiley, New York, 1968.

[58] O. Rodrigues, Sur le nombre de manières de décomposer un polygone en triangles au moyen de diagonales, *J. Math. Pures Appl.* **3** (1838), 547–548.

[59] O. Rodrigues, Sur le nombre de manières d'effectuer un produit de *n* facteurs, *J. Math. Pures Appl.* **3** (1838), 549.

[60] E. Schröder, Vier combinatorische Probleme, *Z. für Math. Phys.* **15** (1870), 361–376.

[61] J. A. Segner, Enumeratio modorum quibus figurae planae rectilinae per diagonales dividuntur in triangula, *Novi Comment. Acad. Sci. Imp. Petropol.* **7** (dated 1758/59, published in 1761), 203–210; available at [55].

[62] N.J.A. Sloane, *A Handbook of Integer Sequences*, Academic Press, New York, 1973.

[63] N.J.A. Sloane and S. Plouffe, *The Encyclopedia of Integer Sequences*, Academic Press, New York, 1995.

[64] R. Stanley, *Enumerative Combinatorics*, vol. 1, 2nd ed., Cambridge University Press, Cambridge, 2012.

[65] R. Stanley, *Enumerative Combinatorics*, vol. 2, Cambridge University Press, New York/Cambridge, 1999.

[66] L. Takács, On the ballot theorems, in *Advances in Combinatorial Methods and Applications to Probability and Statistics*, Birkhäuser, Boston, Massachusetts, 1997, 97–114.

[67] U. Tamm, Olinde Rodrigues and combinatorics, in *Mathematics and Social Utopias in France, Olinde Rodrigues and His Times* (S. Altmann and E. L. Ortiz, eds.), American Mathematical Society and London Mathematical Society, Providence, Rhode Island, pp. 119–129.

[68] H. M. Taylor and R. C. Rowe, Note on a geometrical theorem, *Proc. London Math. Soc.* **13** (1882), 102–106.

[69] H.N.V. Temperley and E. H. Lieb, Relations between the 'percolation' and 'colouring' problem and other graph-theoretical problems associated with regular planar lattices: some exact results for the 'percolation' problem, *Proc. Royal Soc. London, Ser. A* **322** (1971), 251–280.

[70] J. H. van Lint, Combinatorial Theory Seminar, *Lecture Notes in Math.* **382**, Springer, Berlin, 1974, 131 pp.

[71] V. S. Varadarajan, *Euler Through Time: A New Look at Old Themes*, AMS, Providence, Rhode Island, 2006.

[72] W. A. Whitworth, Arrangements of m things of one sort and n things of another sort, under certain conditions of priority, *Messenger of Math.* **8** (1879), 105–114.

[73] B. Ycart, A case of mathematical eponymy: the Vandermonde determinant, *Rev. Histoire Math.* **19** (2013), 43–77.

[74] D. Zeilberger, *Opinion 49* (Oct. 25, 2002); available at `http://tinyurl.com/pzp8bvk`.

Index